Construc

Pocket B

;

The second edition of the *Construction Project Manager's Pocket Book* maintains its coverage of a broad range of project management skills, from technical expertise to leadership, negotiation, team building and communication. However, this new edition has been updated to include:

- revisions to the CDM regulations,
- changes to the standard forms of contract and other documentation used by the project manager,
- the impact of BIM and emerging technologies,
- implications of Brexit on EU public procurement,
- other new procurement trends, and
- ethics and the project manager.

Construction project management activities are tackled in the order they occur on real projects, with reference made to the RIBA Plan of Work throughout. This is the ideal concise reference which no project manager, construction manager, architect or quantity surveyor should be without.

Duncan Cartlidge, FRICS, is a chartered surveyor with extensive experience in the delivery and management of built assets, as well as providing education and training to a wide range of built environment professionals and contractors. He is the author of several best-selling books, including the *Quantity Surveyor's Pocket Book, 3rd edition* and *Estimator's Pocket Book, 2nd edition*, published by Routledge.

Construction Project Manager's Pocket Book

Second Edition

Duncan Cartlidge

LONDON AND NEW YORK

Second edition published 2020
by Routledge
2 Park Square, Milton Park, Abingdon, Oxon, OX14 4RN

and by Routledge
52 Vanderbilt Avenue, New York, NY 10017

Routledge is an imprint of the Taylor & Francis Group, an informa business

First edition published by Routledge 2015

British Library Cataloguing-in-Publication Data
A catalogue record for this book is available from the British Library

Library of Congress Cataloging-in-Publication Data
Names: Cartlidge, Duncan P., author.
Title: Construction project manager's pocket book / Duncan Cartlidge.
Description: Second edition. | Abindgon, Oxon ; New York : Routledge, [2020] |
Series: Routledge pocket books | Includes bibliographical references and index.
Identifiers: LCCN 2019048725 (print) | LCCN 2019048726 (ebook) |
ISBN 9780367437145 (hardback) | ISBN 9780367435936 (paperback) |
ISBN 9781003005216 (ebook)
Subjects: LCSH: Construction industry–Management–Handbooks, manuals, etc. |
Project management–Handbooks, manuals, etc. |
Building–Superintendence–Handbooks, manuals, etc.
Classification: LCC HD9715.A2 C353 2020 (print) | LCC HD9715.A2 (ebook) |
DDC 624.068/4–dc23
LC record available at https://lccn.loc.gov/2019048725
LC ebook record available at https://lccn.loc.gov/2019048726

ISBN: 978-0-367-43714-5 (hbk)
ISBN: 978-0-367-43593-6 (pbk)
ISBN: 978-1-003-00521-6 (ebk)

Typeset in Goudy and Frutiger
by Newgen Publishing UK

For my faithful companion Boris, 2007–2019

Contents

	Preface	ix
1	**Project management – an overview**	1
2	**Pre-construction / RIBA Plan of Work Stages 0–4 / OGC Gateway Stages 1–3C**	70
3	**Construction / RIBA Plan of Work Stage 5**	179
4	**Post-construction / OGC Gateway 4–5 / RIBA Plan of Work Stage 6**	218
5	**Occupancy / RIBA Plan of Work Stage 7**	248
	Appendix A: Financial statement	262
	Appendix B: Practical completion certificate pro-forma for NEC4 form of contract	263
	Appendix C: Final account pro-forma	265
	Appendix D: Design / construction project sample risk list	267
	Appendix E: Sample questionnaire format	269
	Further reading	270
	Glossary	272
	Index	277

Preface

Project management is a comparatively new specialism, having its roots in post-Second World War regeneration, and as such is a discipline that is not just confined to the construction industry.

During the past thirty years or so, project management has been increasingly in demand by construction clients and this is perhaps due to a number of reasons, including the reluctance of architects to take on the role of project manager and the increasing complexity of building and commissioning new and refurbished buildings.

Finding a definition of construction project management is complicated by the use of a variety of similar terms with individuals and / or organisations adopting the title project manager without fully appreciating the nature or the scope of the discipline.

Since the first edition of this pocket book, Building Information Modelling has become mandatory for most public sector projects and there have been major changes to health and safety regulations. In addition, digital construction has arrived and it is anticipated that it will make the role of the project manager more effective.

It is hoped that this pocket book will help to define the role of the construction project manager as well as introducing not only the generic skills required by project managers, but also the specific skills required by the construction project manager.

1

Project management – an overview

In some respects the title of project manager and the term project management are misunderstood and overused in the construction industry, with individuals and / or organisations adopting the title without fully appreciating the nature or the scope of the discipline.

It could be thought that the main attributes of project managers are the so-called hard skills, such as financial analysis, technical know-how, etc., although most project managers and clients consider that effective leadership and the ability to communicate and co-ordinate effectively are equally important. Indeed, recently there has been increased emphasis on the so-called soft skills aspects of project management. This first chapter of the pocket book gives an overview of project management and the role of the project manager as well as outlining the softer (generic) skills required by successful project managers. As will become evident, project management is a global, generic discipline used in many business sectors, of which construction is just one. A criticism of construction project managers is that they have been reluctant to learn from and adopt project management techniques used in other sectors; whether this criticism is warranted is unclear. The remaining chapters of this pocket book relate to project management for construction and development and will be presented with reference to the RIBA Plan of Work (2013) and the OGC Gateway. Although the OGC Gateway model was archived on 22 August 2011 and will not be further developed, it is still widely referred to and for this reason has been included in this handbook.

WHAT IS A PROJECT?

Before it is possible to practise project management it is necessary to define the term project, as distinct from routine day-to-day business activities.

A project can be thought of a temporary group activity designed to produce a unique product, service or result; in the case of construction, a new or refurbished construction project, a new piece of infrastructure, etc. Importantly, a project is temporary, in that it has a defined beginning and end in time, and therefore defined scope and resources. Any activities or processes outside of the project scope are deemed to be 'business as usual' and therefore not part of the project. This transient nature adds pressure to the project manager as it necessitates the development of bespoke solutions. Construction projects traditionally use a management structure known as a temporary multi-organisation, as a project team often includes people who don't usually work together – sometimes from different organisations and across multiple geographies. All must be expertly managed to deliver the on-time, on-budget results, learning and integration that organisations need. In recent times, Latham (1994) and Egan (1998), and subsequently a long list of both public and private sector construction-related reports, urged the introduction of partnering, alliancing and more collaborative working. The construction team has been encouraged to move away from the traditional fragmented approach to delivering projects, but nevertheless the need for project management remains unaltered. Decades after the publication of reports mentioned previously, construction still has a tendency to operate with a silo mentality; overcoming this mentality is a major challenge for construction project managers.

WHAT IS PROJECT MANAGEMENT?

There are a number of definitions of project management which can make pinning down a precise view difficult. This in itself can lead to difficulties, especially when issues of roles and liability are raised. The term project manager is widely used in construction and occurs at many levels in the supply chain. In the UK, management techniques applied to construction, and in particular property development, first started to emerge during the 1970s when a particular approach to property development saw commercial success demanding stricter management and control of time and cost than had previously been the case. During this period contractors began to rebadge themselves as management contractors and some quantity surveyors added project management to their letter heading without fully realising the implications.

Finding a definition of project management in construction is complicated by the use in the industry of a variety of similar commonly used titles such as:

- **Project monitor** – this is distinct from both project management and construction monitoring and is defined in the RICS Project Monitoring Guidance Note as:

  Protecting the client's interests by identifying and advising on the risks associated with acquiring an interest in a development that is not under the client's direct control.

  Project monitoring may include:
  - land and property acquisition,
  - statutory compliance,
  - competency of the developer,
  - financial appraisals,
  - legal agreements,
  - construction costs and programmes, and
  - design and construction quality.

  Some or all of the above are also included in the project manager's brief.

- **Employer's agent** – an employer's agent is employed to administer the conditions of contract, and does not perform the same function as the architect, contract administrator or project manager. For the construction professional the exact position of the employer's agent can be confusing, in particular the duties, if any, that they owe to the contractor. The true employer's agent is a creation of the JCT Design and Build Contract where the contract envisages that the employer's agent undertake the employer's duties on behalf of the employer. Article 3 of the contract gives the employer's agent the full authority to receive and issue:
  - applications,
  - consents,
  - instructions,
  - notices,
  - requests or statements, and
  - otherwise act for the employer.

  The employer's agent has no independent function, but can be thought of as the personification of the employer.

- **Development manager** – as with project manager there are several definitions of the term development manager as defined by:
  - the RICS Development Management Guidance Note,

- CIOB's Code of Practice for Project Management for Construction Development, and
- Construction Industry Council (CIC) Scope of Services 9 (major works).

The RICS Guidance Note defines the role as:

The management of the development process, from the emergence of the initial development concept to the commencement of the tendering process for the construction of the works.

The role of the development manager therefore, may include giving advice on:

- development appraisals,
- planning application process,
- development finance, and
- selection of procurement strategy.

Again, all or some of the above are also encompassed in project management. Some sectors make a definition between the commercial management involving the setting up of the project and the actual implementation and delivery.

According the RICS Project Management Professional Group the most important skills required by construction project managers, as suggested by Young and Duff (1990), are:

- the supervision of others,
- leadership,
- the motivation of others, and
- organisational skills.

Two further terms that require clarification at this stage are:

- **Programme management** – programmes comprise groups of related but interdependent projects and are more concerned with outcomes of strategic benefits, whereas project management concentrates on defined outputs or one-off deliverables.
- **Portfolio management** – refers to the total investment by a client in a variety of projects for the purpose of bringing about strategic business objectives or change.

DEVELOPMENT OF MODERN PROJECT MANAGEMENT

There are some who claim that project management has a long history and was used in the building of the Pyramids 3000 years ago. However, use of techniques such as flogging the workforce at every opportunity can hardly justify the title of motivational project management and for this reason project management is generally thought to have its roots in the nineteenth century.

Three examples of the early pioneers of project management are:

- Frederick Taylor (1856–1915),
- Henry Gantt (1861–1919), and
- William Edwards Deming (1900–1993).

Frederick Taylor

Taylor was born in Germantown, Pennsylvania and in 1878 began working at the Midvale Steel Company where he rose to become foreman of the steel plant and started to apply himself to thoughts about efficiency and productivity. In his book *The Principles of Scientific Management*, Taylor suggests that most managers were ill-equipped to fulfil their role, since they were not trained to analyse and improve work, and seemed incapable of motivating workers. Taylor thought that managers should be able to analyse work (method study) to discover the most efficient way of carrying it out and then should select and train workers to develop their skills in supporting this method. He felt that financial incentives would motivate workers – but that higher productivity would still result in lower wage costs. In fact, he was a strong advocate of co-operation between workers and managers to mutual advantage. Taylor believed strongly in the concept of measurement. By measuring work, and constantly refining and re-measuring working methods, one could work towards an optimal method.

Three fundamental things Taylor taught were:

1. Find the best practice wherever it exists – now referred to as benchmarking.
2. Decompose the task into its constituent elements – now referred to as value management / value engineering.
3. Get rid of things that don't add value – now referred to as supply chain management.

Benchmarking, value engineering and supply chain management are important project management tools which during the past fifty years or so have been adopted, to a greater or lesser extent, by the construction industry and will be referred to again later in this pocket book.

Henry Gantt

Henry Gantt was an associate of Frederick Taylor and is perhaps best known for devising the Gantt chart. Henry Gantt worked as a teacher, draftsman and mechanical engineer before making his mark as an early twentieth-century management consultant. He authored two books on the topic, and he is widely credited with the development of the scheduling and monitoring diagram in the 1910s, now called the Gantt chart, used ubiquitously across industry and manufacturing, which provides easy, visual data on project planning and progress. In fact bar charts were developed 100 years before Gantt and his charts were sophisticated production control tools, not simple representations of activities over time. Throughout his career, Henry Gantt used a wide range of charts; in fact it would be true to say that one of Gantt's core skills was developing charts to display relatively complex data in ways that allowed quick and effective comprehension by managers. However none of these charts were simple forward projections of activities against time (i.e. the conventional 'bar chart' used on modern project management).

William Edwards Deming

William Edwards Deming was an American statistician, college professor, author, lecturer and consultant. Deming is widely credited with improving production in the United States during World War II, although he is perhaps best known for his work in Japan. There, from 1950 onward he taught top management how to improve design (and thus service), product quality, testing and sales (the latter through global markets). Deming made a significant contribution to Japan, becoming renowned for producing innovative high-quality products. Deming is regarded as having had more impact upon Japanese manufacturing and business than any other individual not of Japanese heritage.

Deming was the author of *Out of the Crisis* (1982–1986) and *The New Economics for Industry, Government, Education* (1993), which includes his System of Profound Knowledge and the fourteen Points for Management listed below.

1. Create constancy of purpose for the improvement of product and service, with the aim to become competitive, stay in business and provide jobs.
2. Adopt a new philosophy of co-operation (win-win) in which everybody wins and put it into practice by teaching it to employees, customers and suppliers.
3. Cease dependence on mass inspection to achieve quality. Instead, improve the process and build quality into the product in the first place.
4. End the practice of awarding business on the basis of price tag alone. Instead, minimise total cost in the long run. Move toward a single supplier for any one item, based on a long-term relationship of loyalty and trust.
5. Improve constantly, and forever, the system of production, service and planning of any activity. This will improve quality and productivity and thus constantly decrease costs.
6. Institute training for skills.
7. Adopt and institute leadership for the management of people, recognising their different abilities, capabilities and aspirations. The aim of leadership should be to help people, machines and gadgets do a better job. Leadership of management is in need of overhaul, as well as leadership of production workers.
8. Drive out fear and build trust so that everyone can work more effectively.
9. Break down barriers between departments. Abolish competition and build a win-win system of co-operation within the organisation. People in research, design, sales and production must work as a team to foresee problems of production and use that might be encountered with the product or service.
10. Eliminate slogans, exhortations and targets asking for zero defects or new levels of productivity. Such exhortations only create adversarial relationships, as the bulk of the causes of low quality and low productivity belong to the system and thus lie beyond the power of the workforce.
11. Eliminate numerical goals, numerical quotas and management by objectives. Substitute leadership.
12. Remove barriers that rob people of joy in their work. This will mean abolishing the annual rating or merit system that ranks people and creates competition and conflict.
13. Institute a vigorous programme of education and self-improvement.
14. Put everybody in the company to work to accomplish the transformation. The transformation is everybody's job.

MODERN PROJECT MANAGEMENT TIMELINE

As the above pioneer project managers demonstrate, as a professional discipline project management can realistically be said to have its roots in the late nineteenth century, however project management in its modern form started in the early 1950s, when businesses and other organisations began to see the benefit of organising work around individual projects. This project-centric view of the organisation evolved further as organisations began to understand the critical need for their employees to communicate and collaborate while integrating their work across multiple departments and professions and, in some cases, whole industries. There can be said to be four periods of significant development in project management as outlined below.

1950–1959

During this decade Programme Evaluation Review Technique (PERT) and Construction Project Management systems management were developed by the US military during the development of the Atlas and Polaris ballistic missile programmes and the post-war re-building programme respectively. Both these systems use network techniques with arrows representing activities. The Bechtel Corporation first used the term project management during the construction of a number of large infrastructure projects. The Critical Path Method or arrow diagramming was developed by E.I. du Pont de Numours at Newark, Delaware. Many of the techniques that were to become commonplace during the 1960s were initiated during this time.

1960–1979

The period of mega projects. Following the election of John F. Kennedy as the 35th President of the United States, NASA was charged with getting a man to the moon and back by the end of the 1960s. By the end of the 1970s there was an explosion in the development of management systems and control tools in an attempt to improve the track record for military projects overrunning both in terms of cost and time. Towards the end of the decade there was increased international awareness of the potential of project management. In the US, construction managers were beginning to be used routinely on construction projects. The early 1960s PERT was developed by the US Navy with an emphasis on project events and milestones instead of project activities. The other distinctive characteristic of PERT was the use of probabilistic duration estimates. A report entitled 'A non-computer approach to critical path method for the construction industry' by J.W. Fondahl was published in 1961 utilising the concept of lag values, which came to be known as Precedence Diagramming. A number of professional management bodies were established in Europe and the US. Whereas the 1960s had been dominated by defence / aerospace projects, the 1970s saw the expansion of project-related organisations, typically construction, who started to use project management and project managers as an everyday management function. During this period construction project management was mainly confined to contractors, with owner-driven project management seldom used and only brought into the project after the design stage had been completed. During this period there was an increased recognition that project management was a profession and an increased focus on the refinement of project management tools and techniques.

1980–1994

This period saw the widespread use of IT and its application to project management. Project management matured and degree and master's programmes in project management began to appear in the US. The Project Management Institute developed its Project Management Body of Knowledge (PMBOK) with the first complete edition published in 1986 in the *Project Management Journal*. Examples of major projects undertaken during this period that illustrate the application of high technology and project management tools and practices include:

- The Channel Tunnel project, 1989–1991. This project was an international project that involved two governments, several financial institutions, engineering construction companies and other various organisations from the two countries. The language, use of standard metrics and other communication differences needed to be closely co-ordinated.
- The Space Shuttle Challenger project, 1983–1986. The Challenger Space Shuttle disaster focused attention on risk management, group dynamics and quality management.
- The Calgary Winter Olympics of 1988, which successfully applied project management practices to event management.

1995–present

Until now the emphasis on project management had been on execution and completion stages of projects, but during the early part of this period there was an increasing emphasis being placed on project management at the front end of projects. In addition there was increasing interest in risk, value engineering, etc., with a greater emphasis on project life cycle. This period witnessed the development of project management systems and professional bodies dedicated to training and development of project management and the introduction of project management certification. Latham (1994) calls for construction to learn lessons in project management from other industries.

This period is also dominated by the developments related to the Internet that changed business practices dramatically in the mid-1990s and resulted in the development of Internet and web-based project management applications including the emergence of digital construction.

Also in this period, the Association for Public Management becomes a chartered body and the number of mobile devices and connections exceeds the number of people on the planet.

PROJECT MANAGEMENT GOVERNANCE AND PROFESSIONAL BODIES

Project management practice, standards and education is overseen by several professional bodies.

Association for Project Management (APM) www.apm.org.uk

The APM is a registered charity with over 29,000 individual and 500 corporate members, making it the largest professional body of its kind in Europe. The APM was granted a royal charter in April 2017. The APM's mission statement is *'To provide leadership to the movement of committed organisations and individuals who share our passion for improving project outcomes'*.

The seventh edition of the APM Body of Knowledge (2018) defines the knowledge needed to manage any kind of project. APM qualifications are arranged in four tiers and individual level is assessed by competency assessment as follows:

- **Level A**: Certified Projects Director manages complex project portfolios and programmes.
- **Level B**: Certified Senior Project Manager manages complex projects. Minimum five years of experience.

- **Level C**: Certified Project Manager manages projects of moderate complexity. Minimum three years of experience.
- **Level D**: Certified Project Management Associate applies project management knowledge when working on projects.

Project Management Institute (PMI)
www.pmi.org/uk

The PMI is one of the world's largest associations for project managers with approximately 700,000 members and 550,000 certified practitioners worldwide. The PMI is divided into 300 global chapters over 39 industry sectors. Membership is open to anyone interested in project management on the payment of a modest fee. There are also six project management certification levels including:

- Project Management Professional (PMP),
- PMI Agile Certified Practitioner (PMI-ACP), and
- PMI Risk Management Professional (PMI-RMP).

Competency as a project manager is assessed on experience, education, among other factors.

The Royal Institution of Chartered Surveyors (RICS)
www.rics.org

Unlike the two previous bodies who draw their membership from across a wide range of industrial sectors, the RICS Project Management Professional Group is concerned principally with construction project management. Project management has its own set of competencies and Assessment of Professional Competence (APC) route. There are a number of routes to membership, including an honours degree from an RICS accredited centre, higher degree or via the professional experience route. A number of MSc programmes in Construction Project Management are also available worldwide. According to the RICS the role of the project manager is to lead and motivate the project participants to finish on time, within budget and to meet requirements. This should result in satisfied clients. In simple terms, project management does this by:

- developing the project brief,
- selecting, appointing and co-ordinating the project team,

- representing the client throughout the full development process, and
- managing the inputs from the client, consultants, contractors and other stakeholders.

The RICS also run a number of online e-learning courses which provide technical training against the PMBOK framework.

The Chartered Institute of Building (CIOB)
www.ciob.org

The Building Management Notebook, written and published by the Institute in the early 1960s, was the seminal text for construction management and led the way in reshaping the industry. Originally the Builder's Society, as members work continued to diversify, the Institute changed its name in 1965 to the Institute of Building, later gaining chartered status. Project management in construction and property development also have their roots in the CIOB. An extensive debate within the Institute during the 1980s firmly established project management as a client-orientated discipline. The Code of Practice for Project Management for Construction and Development was first published by the Institute in 1992 and is now in its fifth edition (2014). The Code has made a significant impact on the industry, both in the UK and further afield, and is the premier guide for project management in construction.

CONCEPTS OF ETHICS

Ethical behaviour is developed by people through their physical, emotional and cognitive abilities. People learn ethical behaviour from families, friends, experiences, religious beliefs, educational institutions and media. Business ethics is shaped by societal ethics.

Ethics is a branch of philosophy that covers a whole range of things that have real importance in everyday personal and professional life, including:

- right and wrong,
- rights and duties,
- good and bad,
- what goodness itself is,
- the way to live a good life, and
- how people use the language of right and wrong.

In turn ethics tackles some of the fundamentals of life, for example:

- how should people live, and
- what people should do in particular situations.

Therefore ethics can provide a moral map, a framework, that can be used to find a way through difficult professional issues. Business ethics is about the rightness and wrongness of business practices

Where do ethics come from?

Where do ethics come from; have they been handed down in tablets of stone? Some people do think so and philosophers suggest that ethics originate from:

- God – supernaturalism,
- the intuitive moral sense of human beings – intuitionism,
- the example of 'good' human beings – consequentialism, and
- a desire for the best for people in each unique situation – situation ethics.

ETHICS AND THE PROJECT MANAGER

Ethics, in common with other aspects of construction project management, should be managed and not left to chance.

The Institute of Business Ethics (IBE) was established in 1986 to encourage high standards of business behaviour based on ethical values. The IBE defines business ethics as the application of ethical values (such as fairness, honesty, openness, integrity) to business behaviour, for example:

- Are colleagues treated with dignity and respect?
- Are customers treated fairly?
- Are suppliers paid on time?
- Does the business acknowledge its responsibilities to wider society?

Put simply, business ethics is 'the way business is done around here'.

According to a survey conducted by the IBE and published in 2016, companies regarded the main purpose of having a code of ethical practice as:

1.	Providing guidance to staff	88%
2.	Create a shared and consistent company culture	81%
3.	A public commitment to ethical standards	61%

4. Guarding reputation 27%
5. Reducing operational risk 20%
6. Providing guidance to contractors and the supply chain 20%
7. Helping secure long-term shareholder value 12%
8. Improving the company's competitive position 7%
9. Decreasing the liability in the case of misconduct 2%

Behaving ethically is important and particularly so for construction project managers who operate in a sector that is generally perceived to have low ethical standards. Professions can only survive if the public retains confidence in them. Conducting professional activities in an ethical manner is at the heart of professionalism and the trust that the general public has in professions such as project managers. One of the principal functions for construction-related institutions like the APM, RICS and CIOB is to ensure that their members operate to high ethical standards. For project managers transparency and ethical behaviour is particularly important as they deal on a day-to-day basis with procurement, contractual arrangements, payments, valuations and clients' money.

In a survey carried out by the Chartered Institute of Building in 2013, nearly 40 per cent of those questioned regarded the practice of cover pricing either; '*not very corrupt*' or '*not corrupt at all*', regarding it as the way that the industry operates! In addition, 41 per cent of respondents admitted offering bribes on one or more occasions. One of the major issues from the CIOB survey is a clear lack of definition of corruption and corrupt practices. The industry is one that depends on personal relationships and a particular nebulous area is non-cash gifts that range from pens to free holidays.

The accountancy and business advisory firm BDO's Fraud Track report, which examines all reported fraud over £50,000 in the UK, found that fraud within construction firms soared by almost £6m from £2.6m in 2017 to £8.3m in 2018 to hit an eight-year high, and takes the value of reported fraud in the sector to its highest level since 2010. The increase of fraud in the construction industry bucks the national trend; overall UK-reported fraud more than halved in 2018 – down from £2.1bn to £746.3m (*Construction Manager*, 19 March 2019).

The Bribe Payers Index (Transparency International) illustrates the tendency of industry sectors to indulge in bribery on a scale of 0–10, where 0 indicates the view that companies always bribe and 10 indicates the view that they never do. In the Bribe Payers Index, real estate, utilities and construction languished at the bottom of the league table.

The Global Infrastructure Anti-Corruption Centre has identified thirteen features that make construction particularly prone to corruption, including:

1. Uniqueness – no two construction projects are the same, making comparisons difficult and providing opportunities to inflate costs and conceal bribes.
2. Complex transaction chains – the delivery of infrastructure involves many professional disciplines and tradespeople and numerous contractual relationships that make control measures difficult to implement.
3. Concealment – work is concealed or covered up therefore materials and workmanship are often hidden, e.g. steel reinforcement is cast in concrete, masonry is covered with plaster and cables and pipes enclosed in service ducts.
4. Official bureaucracy – numerous approvals are required from government in the form of licences and permits at various stages of the delivery cycle, each one providing an opportunity for bribery.
5. The scale of infrastructure investments – investments in economic infrastructure such as dams, airports and railways can cost tens of billions of dollars, making it easier to conceal bribes and inflate claims.

Why is ethics important for project managers?

Although the list above appears to be straightforward, things are never quite that simple in practice when matters such as economic survival and competition are added into the mix. The position is even more complicated when operating in countries outside the UK where ideas of ethics may be very different to those expected by the RICS. Ethical behaviour is that which is socially responsible, for example, obeying the law, telling the truth, showing respect for others and protecting the environment. In December 2016 the International Ethics Standards Coalition (IESC), on which the RICS, CIOB and CABE (Chartered Association of Building Engineers) are represented, entered the debate with the publication of 'An ethical framework for the global property market', which contained the following list of ethical principles:

- **Accountability** – practitioners shall take full responsibility for the services they provide, shall recognise and respect client, third-party and stakeholder rights and interests and shall give due attention to social and environmental considerations throughout.

- **Confidentiality** – practitioners shall not disclose any confidential or proprietary information without prior permission, unless such disclosure is required by applicable laws or regulations.
- **Conflict of interest** – practitioners shall make any and all appropriate disclosures in a timely manner before and during the performance of a service. If, after disclosure, a conflict cannot be removed or mitigated, the practitioner shall withdraw from the matter unless the parties affected mutually agree that the practitioner should properly continue. See also; *RICS Professional Standards and Guidance, Global Conflicts of Interest*, 1st edition, March 2017.
- **Financial responsibility** – practitioners shall be truthful, transparent and trustworthy in all their financial dealings.
- **Integrity** – practitioners shall act with honesty and fairness and shall base their professional advice on relevant, valid and objective evidence.
- **Lawfulness** – practitioners shall observe the legal requirements applicable to their discipline for the jurisdictions in which they practice, together with any applicable international laws.
- **Reflection** – practitioners shall regularly reflect on the standards for their discipline and shall continually evaluate the services they provide to ensure that their practice is consistent with evolving ethical principles and professional standards.
- **Standard of service** – practitioners shall only provide services for which they are competent and qualified; shall ensure that any employees or associates assisting in the provision of services have the necessary competence to do so and shall provide reliable professional leadership for their colleagues or teams.
- **Transparency** – practitioners shall be open and accessible, shall not mislead (or attempt to mislead), misinform or withhold information as regards products or terms of service, and shall present relevant documentation or other material in plain and intelligible language.
- **Trust** – practitioners shall uphold their responsibility to promote the reputation of their profession and shall recognise that their practice and conduct bears upon the maintenance of public trust and confidence in the IESC professional organisation and the professions they represent.

ETHICS AND THE LAW

The legislative framework defining fraud has been confused. The principal statutes currently dealing with corruption are the Public Bodies Corrupt Practices Act 1889, the Prevention of Corruption Act 1906 and the Prevention of Corruption Act 1916. This legislation makes bribery a criminal offence whatever the nationality of those involved, if the offer, acceptance or agreement to accept a bribe takes place within the UK's jurisdiction. The Anti-terrorism, Crime and Security Act 2001 has extended UK jurisdiction to corruption offences committed abroad by UK nationals and incorporated bodies. Commercial bribery is currently covered by the Prevention of Corruption Act 1906 insofar as it relates to bribes accepted by agents. The Bribery and Corruption Act 2010 aimed to '*transform the criminal law on bribery, modernising and simplifying existing legislation to allow prosecutors and the courts to deal with bribery more effectively*'. In addition it was hoped that it would also promote and support ethical practice by encouraging businesses to put in place anti-bribery safeguards that ensure all employees are aware of the risks surrounding bribery and that adequate systems exist to manage these. Bribery may include the corruption of a public official as well as commercial bribery, which refers to the corruption of a private individual to gain a commercial or business advantage.

The essential elements of official bribery are:

- giving or receiving
- a thing of value
- to influence
- an official act.

The thing of value is not limited to cash or money. Such things as lavish gifts and entertainment, payment of travel and lodging expenses, payment of credit card bills, 'loans', promises of future employment and interests in businesses can be bribes if they were given or received with the intent to influence or be influenced. The Act makes it a criminal offence to give, promise or offer a bribe and to request, agree to receive or accept a bribe either at home or abroad. The measures also cover bribery of a foreign public official.

The four principal categories of offences are:

1. The offence of bribing another person.
2. Offences relating to being bribed.
3. Bribery of a foreign public official.
4. The new corporate offence – failure to prevent bribery, whereby a commercial organisation (a corporate or a partnership) could be found guilty if a bribe has been made by a person performing services for or on behalf of the commercial organisation with the intention to obtain or retain business or other business advantage for the commercial organisation.

It is a defence for the organisation to show that there were adequate procedures in place designed to prevent employees or agents committing bribery. The penalties on conviction would be the same as for fraud including, in the most serious cases, a sentence of up to ten years' imprisonment following conviction on indictment.

One landmark case to date, in February 2016, resulted in the Sweett Group plc being ordered to pay £2.25 million as a result of a conviction arising from a Serious Fraud Office investigation into its activities in the United Arab Emirates. The company pleaded guilty in December 2015 to a charge of failing to prevent an act of bribery intended to secure and retain a contract with Al Ain Ahlia Insurance Company (AAAI), contrary to Section 7(1)(b) of the Bribery Act 2010, between December 2012 and December 2015. His Honour Judge Beddoe described the offence as a system failure and said that the offending was patently committed over a period of time. This was the first successful conviction under Section 7 of the UK Bribery Act 2010.

UNETHICAL BEHAVIOURS, ACTIVITIES AND POLICIES

Robert Cooke, former Director of the Institute of Business Ethics at DePaul University, has identified fourteen danger signs that an organisation is at risk of unethical behaviour. These are if the organisation:

1. normally emphasises short-term revenues over long-long considerations,
2. routinely ignores or violates internal or professional codes of ethics,
3. always looks for simple solutions to ethical problems and is satisfied with 'quick fixes',
4. is unwilling to take an ethical stand when there is a financial cost to the decision,
5. creates an internal environment that either discourages ethical behaviour or encourages unethical behaviour,
6. usually sends ethical problems to the legal department,
7. looks at ethics solely as a public relations tool to enhance its image,
8. treats its employees differently than its customers,
9. is unfair or arbitrary in its performance-appraisal standards,
10. has no procedures or policies for handling ethical problems,
11. provides no mechanisms for internal whistle-blowing,
12. lacks clear lines of communication within the organisation,
13. is only sensitive to the needs and demands of the shareholders, and
14. encourages people to leave their personal ethical values at the office door.

ETHICS – THE BUSINESS CASE

Does financial wellbeing affect attitudes to ethics – is there a business case for ethics? Do ethics add value? There is evidence (Moore and Robson, 2002) from other sectors that as organisations' turnover increases their social performance worsens! However, businesses and organisations of all sorts – especially the big high-profile ones – are now recognising that there are solid effects and outcomes driving organisational change. There are now real incentives for doing the right thing, and real disincentives for doing the wrong things. As never before, there are huge organisational advantages to behaving ethically, with humanity, compassion and proper consideration for the world beyond the boardroom and the shareholders:

- **Competitive advantage** – customers are increasingly favouring providers and suppliers who demonstrate responsibility and ethical practices. Failure to do so means lost market share, and shrinking popularity, which reduces revenues, profits or whatever other results the organisation seeks to achieve.
- **Better staff attraction and retention** – the best staff want to work for truly responsible and ethical employers. Failing to be a good employer means good staff leave, and reduces the likelihood of attracting good new starters. This pushes up costs and undermines performance and efficiency. Aside from this, good organisations simply can't function without good people.
- **Investment** – fewer investors want to invest in organisations which lack integrity and responsibility because they don't want the association and because they know that for all the other reasons here, performance will eventually decline, and who wants to invest in a lost cause?
- **Morale and culture** – staff who work in a high-integrity, socially responsible, globally considerate organisation are far less prone to stress, attrition and dissatisfaction. Therefore they are happier and more productive. Happy productive people are a common feature in highly successful organisations. Stressed, unhappy staff are less productive, take more time off, need more managing and also take no interest in sorting out the organisation's failings when the whole thing implodes.
- **Reputation** – it takes years, decades, to build organisational reputation, but only one scandal to destroy it. Ethically responsible organisations are far less prone to scandals and disasters. And if one does occur, an ethically responsible organisation will automatically know how to deal with it quickly, openly and honestly. People tend to forgive

organisations who are genuinely trying to do the right thing. People do not forgive, and are actually deeply insulted by, organisations who fail and then fail again by not addressing the problem and the root cause. Arrogant leaders share this delusion that no-one can see what they're up to. Years ago maybe they could hide, but now there's absolutely no hiding place.

- **Legal and regulatory reasons** – soon there'll be no choice anyway; all organisations will have to comply with proper ethical and socially responsible standards. And these standards and compliance mechanisms will be global. Welcome to the age of transparency and accountability. So it makes sense to change before you are forced to.
- **Legacy** – even the most deluded leaders will admit in the cold light of day that they'd prefer to be remembered for doing something good, rather than making a pile of money or building a great big empire. It's human nature to be good. Humankind would not have survived were this not so. The greedy and the deluded have traditionally been able to persist with unethical irresponsible behaviour because there's been nothing much stopping them, or reminding them that maybe there is another way. But no longer. Part of the reshaping of attitudes and expectations is that making a pile of money and building a great big empire are becoming stigmatised. What's so great about leaving behind a pile of money or a great big empire if it's been at the cost of others' wellbeing or the health of the planet? The ethics and responsibility zeitgeist is fundamentally changing the view of what a lifetime legacy should and can be. And this will change the deeper aspirations of leaders, present and future, who can now see more clearly what a real legacy is.

ETHICAL DECISION-MAKING MODELS

The IBE has developed a nine-step model for developing and implementing a code of ethics:

1. Understand your context – at the outset, it is necessary to understand the catalyst for the code. Knowledge of this will help create an appropriate framework to develop your code.
2. Establish board-level support – corporate values, ethics and culture are matters of governance. Without senior leadership endorsement, embedding a code of business ethics is unlikely to be effective.

3. Articulate your core (ethical) values – ensure that your organisation has articulated its core values. Ethical values in particular should form the basis of the code and guide employee decision-making through ethical dilemmas.
4. Find out what bothers people – it is important to find out what issues are of particular relevance to your employees and sector, so copying the code of another organisation is not the way to proceed.
5. Choose your approach – codes can be structured according to stakeholders, issues, values or in a 'hybrid' way. Each organisation will need to choose which model best suits its individual requirements.
6. Draft your code – the drafting of the code should produce a document which is clear, inclusive and accessible. It should be principles-based, give guidance to staff and detail who to ask when unsure of the right way forward.
7. Test it – to ensure your code is fit for purpose, it needs to be piloted with a cross-section of employees drawn from different locations and levels of employment.
8. Launch it – the launch event of a new or revised code should be memorable, engage employees and raise their awareness of the importance of doing business ethically.
9. Monitor it – the launch of the code is just the beginning of the journey. Ongoing monitoring, training on its use and rewarding those who demonstrate ethical leadership are also required.

Other commonly used ethical decision-making models are:

- American Accounting Association model,
- Laura Nash model,
- Tucker's 5-question model,
- Mary Guy model,
- Rion model, and
- Langenderfer and Rockness model.

DEFINING THE ROLE

As discussed previously, project management has many definitions, even being referred to as professional art rather than technical management. For the purposes of this pocket book, project management may be regarded as the professional discipline that ensures that the management function of project delivery remains separate from the design / execution functions of a project. Confusingly, there has been a tendency during the last thirty years or so for quantity surveyors working in private practice to call themselves project managers in order to differentiate themselves from construction quantity surveyors, without fully appreciating the breadth of the discipline they aspire to.

Typically, project managers will be appointed at the beginning of a project and will assist the client in developing the project brief and then selecting, appointing and co-ordinating the project team. The project manager will usually represent the client throughout the full development process; managing the inputs from the client, consultants, contractors and other stakeholders.

The activities that are most commonly involved with construction project management are described in Chapter 2.

Project management is all about setting and achieving reasonable and attainable goals. It is the process of planning, organising, and overseeing how and when these goals are met. Unlike business managers who oversee a specific functional business area, project managers orchestrate all aspects of time-limited, discrete projects.

During the 1980s the Ethics, Standards and Accreditation project of the PMI established the three constraints of project management; time, cost and quality. In addition to project time and cost management, a third function – quality – was added, to be followed eventually by a fourth – scope – as illustrated in Figure 1.1. To the above some project managers add a fifth constraint – risk.

Constraint – scope

It is important from the outset that all members of the project team are clear about the scope of the project. It could be defined as 'the construction of a mixed-use development comprising retail units, commercial space and residential accommodation and associated external works and parking'. However, the scope can be further clarified by defining the 'what' of the project as follows:

- What will you have at the end of the project?
- What other deliverables could sensibly be carried out at the same time?
- What (if anything) is specifically excluded from the project?
- What are the gaps or interaction (if any) with other projects?

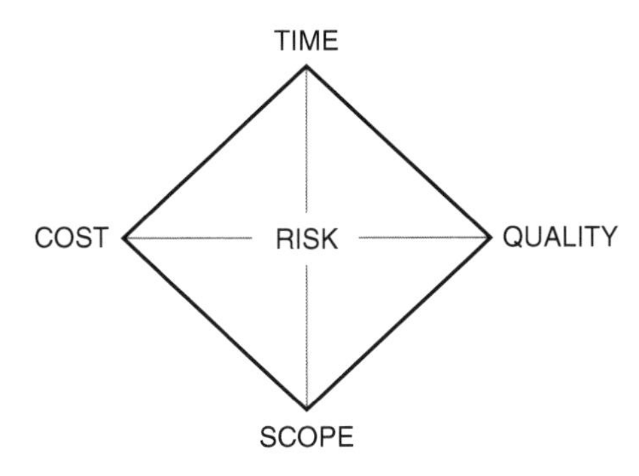

Figure 1.1 Project constraints

- What is the chance the scope of the project will creep?
- What assumptions have to be made?
- What significant difficulties have to be overcome?
- What specific conditions or constraints have been stipulated by the client?

Constraint – cost

The cost constraint could be defined in terms of the cost limit or budget for the project.

Constraint – time

The time constraints could be defined as the time to complete the project from access to the site as entered into the contract.

Constraint – quality

The project should result in a functionally efficient building. Quality is all about the extent that something is fit for the purpose for which it is intended. Value engineering can be used to help achieve this. A problem when defining quality is agreeing within the team what is fit for purpose; it has been known for a project manager to disagree with a client over what constitutes fit for purpose.

Constraint – risk

Monitor the progress of the project according to the project plan and the above variables, deal with issues as they arise during the project, look for opportunities to reduce costs and speed up delivery time, and plan, delegate, monitor and control.

GENERIC / SOFT PROJECT MANAGEMENT SKILLS

Project management and the project manager are not unique to the construction industry and there are a number of generic project management skills common to all sectors and industries, for example:

- leadership,
- motivation,
- communication, and
- budgetary control.

Construction projects, more so than projects in other sectors, take place in a wider geographical, political and regulatory environment (Figure 1.2), and these aspects of construction project management are ignored at the project manager's peril.

A construction project is often part of a larger programme of works, for example a project that is part of a large urban regeneration scheme, and therefore it is true to say that construction project management requires a unique combination of tools and techniques, and these will be fully discussed and explained later in this pocket book.

Project management, then, is the application of knowledge, skills and techniques to execute projects effectively and efficiently. It's a strategic competency for organisations, enabling them to tie project results to business goals – and therefore, compete more effectively and efficiently in their markets. A survey conducted by El-Sabaa (2001) attempted to measure the

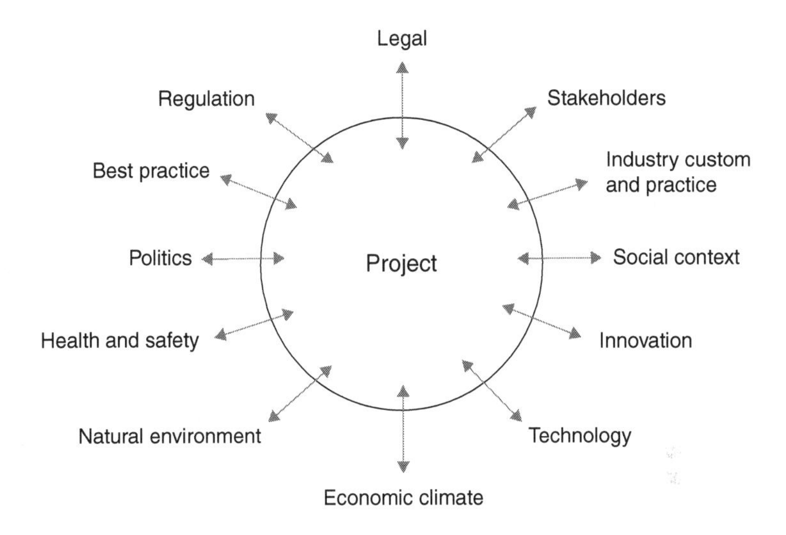

Figure 1.2 Interaction between the immediate and wider project environments

Source: Based on BSi PD 6079-4:2006.

importance to successful project management in the agriculture, electricity and IT sectors of:

- **human skills** – the ability to communicate with and motivate people,
- **organisational skills** – management of time, information and costs), and
- **technical skills** – industry-specific knowledge and expertise.

The survey revealed, as illustrated in Figure 1.3 that, regardless of the sector, human and organisational skills were more highly rated project manager attributes than technical know-how. Project managers can come from a variety of backgrounds and disciplines but need to have the skill set and competencies to manage all the aspects of a wide range of projects and personnel from initial brief to handover and use.

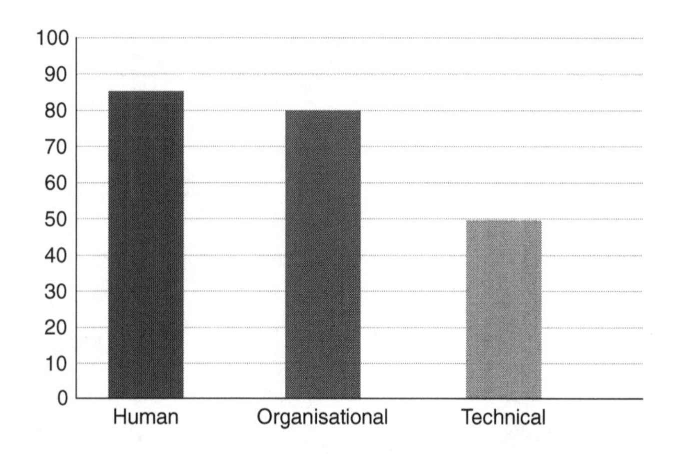

Figure 1.3 Importance of project management skills across industries

Source: El-Sabaa (2001).

Leadership

Project managers must be able to work with a variety of people with differing skill sets. In order to facilitate and establish effective leadership the following points should be considered:

- Good communications – if the project manager is too technical, they may lose people; too basic and may be perceived as being patronising.
- Keeping tabs on external suppliers' resources and choosing carefully.
- Understanding team working – clarify / define the following concepts:
    - **Authority** – the ability to make binding decisions concerning the project. It defines decisions the project manager is able to take but does not define what results have to be achieved.
    - **Responsibility** – having to deliver specific activities or outcomes – for example producing a cost plan by Friday. Unlike authority, responsibility defines the results but is unconcerned about the decisions necessary to obtain the results.
    - **Accountability** – ensuring awareness that actions or lack of actions attract corresponding consequences.

There are a variety of leadership styles that have been identified over the past hundred years or so, but this chapter will concentrate on three, namely:

- autocratic (authoritarian),
- participative (democratic), and
- delegative.

A good project manager may use all three styles, depending on what forces are involved between the team members and the project manager, and the situation.

Autocratic leadership

Autocratic leaders, also known as authoritarian leaders, provide clear expectations for what needs to be done, when it should be done and how it should be done. There is also a clear division between the leader and the followers. Autocratic leaders make decisions independently, with little or no input from the rest of the group. It has been found that decision-making is less creative under autocratic leadership and that often it is more difficult to move from an autocratic style to a democratic style than vice versa. Abuse of this style is usually viewed as controlling, bossy and dictatorial. The autocratic / authoritarian approach includes being arbitrary, controlling, power-oriented and close-minded. The cluster has often been described in pejorative terms. It means taking full and sole responsibility for decision and control of followers' performance. This style of leadership is best applied to situations where there is little time for group decision-making or where the leader is the most knowledgeable member of the group. This approach demands obedience, loyalty and strict adherence to roles in order that decisions are carried out.

Participative leadership

Participative leadership, also known as democratic leadership, is generally the most effective leadership style. Participative leaders offer guidance to group members, but they also participate in the group and allow input from other group members. Participative leaders encourage group members to participate, but retain the final say over the decision-making process. Group members feel engaged in the process and are more motivated and creative.

Delegative leadership

It will not be possible for the project manager to deal personally with every detail of the project and therefore it may, depending on the size and complexity of the project, be necessary to delegate authority for some aspects of the day-to-day business. While ultimate responsibility cannot be relinquished and remains with the project manager, delegation of authority carries with it the imposition of a measure of responsibility. The extent of the authority delegated must be clearly stated to the team member and when deciding which tasks to delegate the following should be considered:

- Retain tasks that you do best.
- Recognise that others may do a better job and have more expertise for some tasks.
- Don't delegate tasks that cannot be clearly defined.
- As project manager, try not to retain tasks that are on the critical path as it is almost certain that at some time during the project you may not have sufficient time to devote to these important operations.
- Ensure that the person delegated is fully briefed as to what has to be done and what the deadlines are.
- Ensure that time is allocated to monitor the progress of the tasks that have been delegated; don't assume that matters are progressing well.

It has been found that group members under delegative leadership, also known as laissez-faire leadership, were the least productive of all three groups discussed here. The members in this group also made more demands on the leader, showed little co-operation and were unable to work independently. Delegative leaders offer little or no guidance to group members and leave decision-making up to group members. While this style can be effective in situations where group members are highly qualified in an area of expertise, it often leads to poorly defined roles and a lack of motivation. The democratic or egalitarian leadership approach reflects concern about the team members in many different ways.

Leadership should be considerate, democratic, consultative, participative and employee-centred; concerned with people and maintenance of good working relations; supportive and oriented toward facilitating interaction; relationship oriented, and oriented toward group decision-making.

The correct approach

A good project manager uses all three leadership styles, depending on what forces are involved between the followers, the leader and the situation. Some examples include:

- Using an autocratic style on new team members who are new to the process. The project manager is competent and a good coach. The team members are motivated to learn a new skill and the situation is a new environment for the team.
- Using a participative style with a team who know their job. The leader knows the problem, but does not have all the information. The team members know their jobs and want to become part of the team.
- Using a delegative style with a team member who knows more about the job than you. Project managers cannot do everything and the team members need to take ownership of their job! In addition, this allows the project manager to be more productive.
- Using all three:
  - telling the team members that a procedure is not working correctly and a new one must be established (autocratic);
  - asking for their ideas and input on creating a new procedure (participative);
  - delegating tasks in order to implement the new procedure (delegative).

Forces that influence which style should be used by the project manager include:

- the amount of time available,
- whether relationships are based on respect and trust or on disrespect,
- who has the information (you, the team members, or both?),
- how well the team members are trained and how well you know the task,
- internal conflicts,
- type of task (structured, unstructured, complicated, or simple?), and
- regulations or established procedures.

Motivation

A basic perspective of motivation looks something like this:

A well-motivated team will obviously be more productive, and their workplace generally a happier place to be and work in than one that lacks motivation. There have been many studies into what motivates human beings and successful project managers should be aware of at least some of the research that has been carried out in this field. Perhaps one of the most widely referred to theories is one first published by Abraham Maslow in 1954 entitled, *Motivation and Personality*.

Classifying needs

People have different needs (Figure 1.4), which means that project managers have to try to understand the whole gamut of needs and who has them, in order to begin to understand how to design teams that maximise productivity. Part of what a theory of motivation tries to do is explain and predict who has which needs, which can turn out to be exceedingly difficult.

The idea is that people start at the base of the pyramid, where the needs are more urgent, and work upwards:

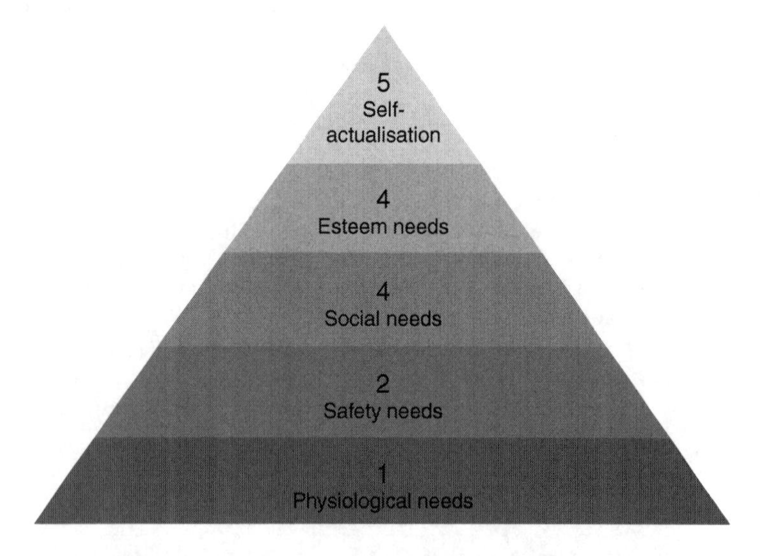

Figure 1.4 Applying the Maslow theory to project management

Step 1 – Physiological needs	Salary, decent working environment
Step 2 – Safety needs	Safe working conditions, job security
Step 3 – Social needs	Good team atmosphere, friendly supervision
Step 4 – Esteem needs	Impressive job title, recognition of achievements
Step 5 – Self-actualisation	Opportunities for creativity and personal growth, promotion

The model is sometimes criticised as only applying to middle-class workers!

Specific examples of these types are given in Table 1.1, in the work context.

Communication

Good communication is central to a project manager's role and needs careful thought and planning, particularly in the digital age when information travels at the click of a mouse – including items that haven't been carefully considered. Breakdown in communications is often cited as one of the principal reasons for project failures.

Table 1.1 Classification of needs

Need	*Manifestation*
Self-actualisation	Training, advancement, growth, creativity
Esteem	Recognition, high status, responsibilities
Belongingness	Teams, depts., fellow workers, clients, supervisors, subordinates
Safety	Work safety, job security
Physiological	Salary

Types of communication

Project managers use a variety of ways to communicate; these include informal face-to-face meetings, phone calls, email and meetings in a formal setting. It is the final communication vehicle on the list, meetings, that attracts widespread criticism from many project teams who would claim to spend many hours in a succession of non-productive meetings.

Running a productive meeting

There are a number of golden rules for running meetings productively and effectively (see also Chapter 3):

1. Prepare in advance by deciding what the objectives of the meeting are and convert these objectives into an agenda and circulate this in advance, together with any necessary paperwork of back-up documents. Some project managers set timed agendas, where a period of time is set against each item. One of the main benefits of a timed agenda (other than time efficiency) is that it forces the participants to plan for the meeting in advance. This keeps unnecessary topics from taking up time and focus. It also forces the group to prioritise to fit within the time constraints.
2. Make sure that the right people attend the meeting, for example people who are able to make decisions without reference to others, or people with a particular expertise.
3. Select a venue that is going to conducive to productive interaction and ensure that equipment such as whiteboards and projectors are available.

Conversation

A project manager should seek to communicate rather than chatter. To ensure an efficient and effective conversation, there are three considerations:

- make the message understood,
- receive / understand the intended message sent to you, and
- exert some control over the flow of the communication.

It is also important to learn to listen as well as to speak. If you do not explicitly develop the skill of listening, you may not hear the suggestion / information.

AVOID AMBIGUITY

As a project manager your view of words should be pragmatic rather than philosophical. Thus, words mean not what the dictionary says they do but rather what the speaker intended.

HAVE A PURPOSE

As with all effective communication, the purpose of the conversation and the plan for achieving it should be decided on in advance. Some people are proficient at thinking on their feet – but this is generally because they already have clear understanding of the context and their own goals.

ASSERTIVENESS

If, as occasionally happens, someone starts an argument or even loses their temper, the best policy is to be quietly assertive. Much has been written to preach this simple fact and commonly the final message is a three-fold plan of action:

- Acknowledge what is being said by showing an understanding of the position, or by simply replaying it – a polite way of saying 'I heard you already'.
- State your own point of view clearly and concisely with perhaps a little supporting evidence.
- State what is to happen next – move the agenda forward.

There will certainly be times when a bit of quiet force will win the day but there will be times when this will get nowhere, particularly with more senior and unenlightened management. In the latter case, try to agree to abide by the decision of the senior manager but you should make your objection with reasons clearly known. However, always be aware that junior members of the team might be right when they disagree with you and if events prove them so, acknowledge that fact gracefully.

CONFRONTATIONS

When faced with a difficult situation with a team member, be professional and try not to lose self-control. Some project managers believe it is useful for discipline to keep staff a little nervous. These managers are slightly volatile and will be willing let rip when the situation demands. If this approach is adopted then the project manager must try to be consistent and fair so that team members know where they stand. Remember that insults and name-calling are ineffective as people are unlikely to actually listen to what you have to say; in the short term it may be a relief at getting it off your chest, but in the long run the problem is simply perpetuated as the root cause of the problem is not addressed. Before responding, stop, establish the desired outcome, plan how to achieve this and then speak. Finally, if criticism of a team member is justified, always assume that there has been a misunderstanding of the situation; ask questions first and check the facts – this may save much embarrassment.

SEEKING INFORMATION

There are two ways of asking a question:

- The closed question – far easier for the respondent to be evasive.
- The open question – compels the respondent to be more informative.

Imagine that at a project meeting you ask a team member about the progress of a report along these lines:

Q: Is the client report finished now?
A: Yes, more or less.

Q: Is there much left to do?
A: Just bits and pieces.

Q: Will that take long?
A: No, not really.

In the above example the questions are not helping the project manager to get an accurate picture of the status of the report. However, if the questions are asked as open questions, they are more likely to elicit an informative response:

Q: What do you have to do to finish the client's report?

Q: When will the report be finished?

It is less easy to be evasive if the question is started with; *what*, *when*, *why*, *where*, *how*, etc.

LETTING OTHERS SPEAK

Of course, there is more to a conversation, managed or otherwise, than the flow of information. The project manager may also have to gain information by winning the attention and confidence of the other person. To get a team member to give you all their knowledge, you must give them all your attention; talk to them about their view on the subject. Ask questions: What do you think about that idea? Have you ever met this problem before? How would you tackle this situation?

Silence is also very effective – and much under-used. People are nervous of silence and try to fill it. A project manager can use this when seeking information. Ask the question, lean back, the person answers, nod and smile, keep quiet – the person will continue with more detail simply to fill your silence.

TO FINISH

At the end of a conversation people should have a clear understanding of the outcome. For instance, if there has been a decision, restate it clearly in terms of what should happen and by when; summarise the significant aspects of what has been learned.

Project managers need to communicate to co-ordinate their own work and that of others. Without explicit effort a conversation will lack communication and so the work too will collapse though misunderstanding and error. The key is to treat a conversation as any other managed activity: by establishing an aim, planning what to do and checking afterwards that the aim has been achieved. Only in this way can the project manager work effectively with others in building through common effort.

Budgetary control

Budgetary control is the process of developing a spending plan and periodically comparing actual expenditures against that plan to determine if it or the spending patterns need adjustment to stay on track. This process is necessary to control spending and meet various financial goals. Both the public and private sectors rely heavily on budgetary control to manage their spending activities. The first step in budgetary control involves defining the scope of the project or programme and developing detailed cost estimates. From this follows the creation of a budget – a document detailing how much money can be dedicated to different aspects of the project, based on projected expenses and income (a financial road map) – and using the budget as a baseline, work can begin. In construction, materials costs might rise beyond the inflation accounted for in the original budget, creating a cost overrun. Conversely, a company might be able to save money on part of a project because it costs less than originally expected. All variations are noted and discussed. If they become extreme, budgetary control measures may come into play.

In some cases, adjustments to spending behaviour may not be possible. Instead, a revised budget is necessary. Revisions may reveal the need for additional funding, forcing parties in charge of budgetary control to discover where that money will come from. This could include taking on debt, cutting the scope of a project or moving funds over from another project or programme to keep it going. A company, for example, could partially remove funds from a department to push through completion of an important project.

CHANGE MANAGEMENT

Very often new projects involve change, either in terms of the organisation or to personnel within the organisation, and the project manager should be aware of this. Project management and change management are distinct but interwoven techniques.

As previously defined, project management is the application of knowledge, skills and techniques to execute projects effectively and efficiently, whereas change management refers to the process, tools and techniques to manage the people side of change to achieve the required business outcome.

On occasions a separate change manager may be appointed, although in practical terms change managers and project managers understand each other's discipline and share critical common elements and therefore

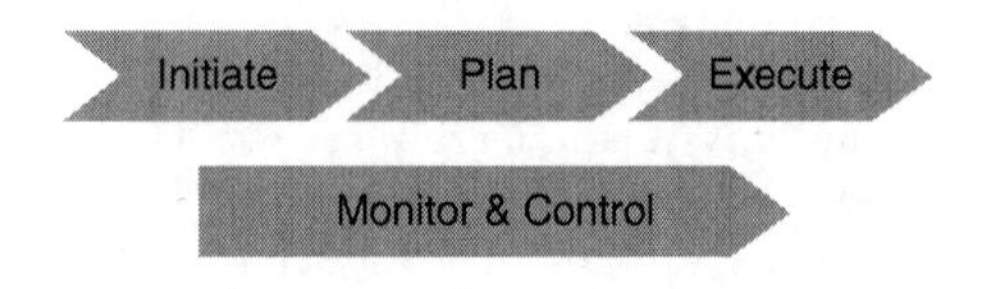

Figure 1.5 Project management life cycle

Figure 1.6 Change management life cycle

the project manager could take change management under their wing. Figures 1.5 and 1.6 compare the two disciplines.

Change management incorporates the organisational tools that can be utilised to help organisations and individuals make successful personal transitions resulting in the adoption and realisation of change.

Steps in the change management process are said to be (see Figure 1.6):

- planning for change,
- managing change and
- reinforcing change.

In the widest sense, change management is a structural approach for moving organisations from their current state to a future state with anticipated business and organisational benefits. It helps organisations to adapt and align to new and emerging market forces and conditions. Delivery and handover of a successful project may well involve organisational change. In order to get the maximum benefit from a project a well-managed handover is essential and project managers should be able to manage the process successfully.

The steps for an effective change management process in project management are:

- Formulate the change by identifying and clarifying the need for change and establishing the scope of change.
- Plan the change by defining the change approach and planning stakeholder engagement as well as transition and integration.
- Implement the change by preparing the organisation for change, mobilising the stakeholders and delivering project outputs.
- Manage the change transition by transitioning the outputs into business operations, measuring the adoption rate and the change outcomes and benefits and adjusting the plan to address discrepancies.
- Sustain the change on an ongoing basis through communication, consultation and representation of the stakeholders, conducting sense-making activities and measuring benefits.

A project manager can influence the culture to change by:

- assessing stakeholder change resistance and / or support for change,
- ensuring clarity of vision and values among stakeholders,
- creating an understanding among the various stakeholder groups about their individual and interdependent roles in attaining the goals of the change initiative, and
- building strong alignment between stakeholder attitudes and strategic goals and objectives.

CHANGE MANAGEMENT MODELS

Lewin's Change Management Model

This change management model was created in the 1950s by psychologist Kurt Lewin. Lewin noted that the majority of people tend to prefer and operate within certain zones of safety. He recognised three stages of change:

- **Unfreeze** – most people make a conscious effort to resist change. In order to overcome this tendency, a period of thawing or unfreezing must be initiated through motivation.
- **Transition** – once change is initiated, the company moves into a transition period, which may last for some time. Adequate leadership and reassurance is necessary for the process to be successful.
- **Refreeze** – after change has been accepted and successfully implemented, the company becomes stable again, and staff refreeze as they operate under the new guidelines. While this change management model remains widely used today, it takes time to implement. Since it is easy to use, most companies tend to prefer this model to enact major changes.

McKinsey 7-S Model

The McKinsey 7-S model offers a holistic approach to organisation. This model, created by Robert Waterman, Tom Peters, Richard Pascale and Anthony Athos during a meeting in 1978, has seven factors that operate as collective agent of change:

1. Shared values
2. Strategy
3. Structure

4. Systems
5. Style
6. Staff
7. Skills.

The McKinsey 7-S Model offers four primary benefits:

- It offers an effective method to diagnose and understand an organisation.
- It provides guidance in organisational change.
- It combines rational and emotional components.
- All parts are integral and must be addressed in a unified manner.

The disadvantages of the McKinsey 7-S Model are as follows:

- When one part changes, all parts change, because all factors are interrelated.
- Differences are ignored.
- The model is complex and companies using this model have been known to have a higher incidence of failure.

Kotter's 8 Step Change Model

Created by Harvard University Professor John Kotter, this model causes change to become a campaign. Employees buy into the change after leaders convince them of the urgent need for change to occur. There are eight steps involved in this model:

1. Increase the urgency for change.
2. Build a team dedicated to change.
3. Create the vision for change.
4. Communicate the need for change.
5. Empower staff with the ability to change.
6. Create short-term goals.
7. Stay persistent.
8. Make the change permanent.

Significant advantages to the model are as follows:

- The process is an easy step-by-step model.
- The focus is on preparing and accepting change, not the actual change.
- Transition is easier with this model.

However, there are some disadvantages offered by this model:

- Steps can't be skipped.
- The process takes a great deal of time.
- It doesn't matter if the proposed changed is a change in the process of project planning or general operations.

Adjusting to change is difficult for an organisation and its employees and using almost any model is helpful to the project manager, as it offers leaders a guideline to follow, along with the ability to determine expected results.

ORGANISATIONAL DEVELOPMENT

Organisational development is a technique to formalise approaches of organisations that are subject to continuous and rapid change. Ways of implementing organisational development include:

- employing external consultants to advise on change,
- establishing an internal department to instigate organisational change, and
- integrating the change process within the mainstream activities of the organisation.

There are a variety of opinions as to which approach is best as each has its strengths and weaknesses.

BUSINESS PROCESS RE-ENGINEERING

Business process re-engineering on face value sounds very similar to organisational development, and in practice the two approaches can be difficult to separate

The idea of re-engineering was first propounded in an article in *Harvard Business Review* in July–August 1990 by Michael Hammer, then a professor of computer science at MIT. The method was popularly referred to as business process re-engineering (BPR), and was based on an examination of the way information technology was affecting business processes. BPR promised a novel approach to corporate change, and was described by its inventors as a "fundamental rethinking and radical redesign of business processes to achieve dramatic improvements in critical measures of performance such as cost, quality, service and speed".

The technique involves analysing a company's central processes and reassembling them in a more efficient fashion and in a way that overrides long-established and frequently irrelevant functional distinctions – a similar approach as that adopted by value engineering. Throughout this pocket book there is frequent reference to the traditional silo mentality of the construction industry; silos that are often protective of information, for instance, and of their own position in the scheme of things. Breaking up and redistributing the silos into their different processes and then reassembling them in a less vertical fashion exposes excess fat and forces organisations to look at new ways to streamline themselves.

One of the faults of the idea, which the creators themselves acknowledged, was that re-engineering became something that managers were only too happy to impose on others but not on themselves. Hammer's follow-up book was pointedly called *Reengineering Management*. 'If their jobs and styles are left largely intact, managers will eventually undermine the very structure of their rebuilt enterprises,' he wrote with considerable foresight in 1994. BPR has been implemented with considerable success by some high-profile organisations, however it has been suggested that construction, due to its fragmented nature, is a barrier to inter-organisational change.

PROJECT MANAGEMENT TOOLS AND TECHNIQUES

The widespread use of programmes and IT packages during the past thirty years or so has revolutionised the way in which project managers work. Systems such as:

- PRINCE2,
- PRIME,
- Microsoft Project, and
- Newforma

are now widely used and the increasing adoption of Building Information Modelling (BIM) help the project manager work more efficiently and effectively.

PRINCE2

PRINCE2 (or PRojects IN a Controlled Environment) is a project methodology developed by the private sector and adapted for use in the public sector, originally for use on IT projects. The system is not a software package but can be used on a range of projects from small individual ones to mega projects. Although not in itself a software package there are over fifty tools supporting the methodology.

However, it is not a standard approach and needs to be customised for each project. PRINCE is open access, that is to say free, and is used throughout the UK as well as internationally, although it will be necessary to invest in training to get the most out of the system or at least buy the official PRINCE2 book bundle from the OGC / Cabinet Office for about £85 from www.itgovernance.co.uk/prince2.aspx. Foundation training courses are available for approximately £300. At the last count there were more than 250,000 certified project managers who had passed the PRINCE2 practitioners examination worldwide.

As Figure 1.7 illustrates, PRINCE2 is an integrated framework of processes and themes that addresses the planning delegation, monitoring and control of six aspects of project management. PRINCE2 uses four integrated elements:

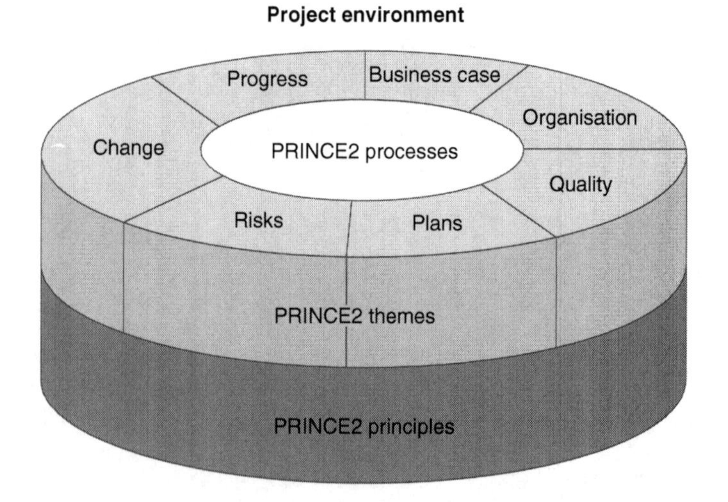

Figure 1.7 PRINCE2 integrated framework

Source: © APM Group Ltd.

1. Seven principles – best practice and good characteristics:
   (a) continued business justification,
   (b) learn from experience – previous projects,
   (c) defined roles and responsibilities – accountability and responsibility,
   (d) manage by stages – break big projects into smaller chunks,
   (e) manage by exception – authority delegation technique,
   (f) focus on projects – define product before development, and
   (g) tailor PRINCE2 to suit the project – customising.
2. Themes – items that need to be continually assessed during the project and customised to suit the project:
   (a) business case – one of the main drivers,
   (b) organisation – who, responsibilities, communication,
   (c) quality – define level of quality, controlling quality,
   (d) plans – approach, resources,
   (e) risk – what if …?
   (f) change – what is the impact of change?
   (g) progress – where are we know, where are we going, should we continue?
3. Processes:
   (a) starting a project – feasibility; sketching out and looking at the project to decide whether it will work,
   (b) initiating a project – the business case, risk register and sensitivity analysis / resources / security / legal – compliance / technology,
   (c) directing a project – setting key decision points,
   (d) controlling a stage – day-to-day activities controlled by the project manager,
   (e) managing product delivery,
   (f) managing a stage boundary – report on stage and plan next one, and
   (g) closing a project.
4. Tailoring PRINCE2:
   (a) PRINCE2 should be customised to individual projects.

PRINCE2 identifies six project variables or performance targets:

- **Time** – when will the project finish?
- **Cost** – are we within budget?
- **Quality** – is it fit for purpose?
- **Scope** – how can we avoid scope creep / uncontrolled change?
- **Benefits** – why are we doing this project?
- **Risk** – risk management what happens if…?

Another feature of the system are targets for these variables. These targets are set in at the planning stage and are regularly checked by the project manager during the project.

Benefits of using PRINCE2

- Proven best practice – used for over thirty years.
- It is flexible and can be applied to any project.
- It recognises project responsibilities, accountability and roles.
- There is product focus that is well defined at the outcome.
- It brings in managers at key moments.
- The viability of the business case is constantly reviewed.
- It integrates risk management into routine project management

PRIME

Unlike PRINCE2, PRIME is ISO 21500:2012 (currently under review) compliant. Initially free, a fee is now payable according to the size of an organisation. PRIME is a standard project management methodology. It is claimed that training is not required.

Microsoft Project

Microsoft Project (2016) is a project management software program, developed and sold by Microsoft, which is designed to assist a project manager in developing a plan, assigning resources to tasks, tracking progress, managing the budget and analysing workloads. Microsoft Project was the company's third Microsoft Windows-based application, and within a couple of years of its introduction it became the dominant PC-based project management software.

Newforma

In addition to project management frameworks there are also a number of other solutions available that can both enable greater efficiency and collaboration between members of the project team. One such solution is Newforma.

Newforma is a system that facilitates project collaboration using the Project Cloud; a web-hosted construction collaboration software that integrates information from the design, construction and owner's team that can be accessed from portable mobile devices. Project Information Management

(PIM) addresses the basic needs of organising, finding, tracking, sharing, monitoring and reusing technical project information and communications in a way that is completely aligned with the people and processes that need the information.

PROJECT MANAGEMENT PHASES

Generally, the project management process falls into five stages:

1. initiating,
2. planning / organisation,
3. executing / implementation,
4. monitoring and controlling, and
5. closing / evaluation.

Although perhaps a negative place at which to start an analysis of project management skills, it is a sad fact that many projects fail to achieve a satisfactory conclusion in the construction industry and therefore it is essential to be aware of the potential pitfalls:

1. Cost overruns – historically the construction industry has a poor reputation for delivering projects on budget. This is usually passed off with statements such as 'every building project is bespoke and carried out in conditions (adverse weather, for example) which makes hitting budget very difficult'.
2. Unrealistic programmes / schedules – along with cost overruns, the other curse of construction projects is finishing late. This again is often explained away by the excuses given for cost overruns.
3. Not meeting the expectations of the client – sub-optimal project performance is common in construction projects when the client's perception and the design team's perception of the finished project differ. This can be down to just plain arrogance on behalf of the client's professional advisors, thinking that they know better than the client, failures to understand the need for the project or problems in the brief process.
4. Failure to clearly define the scope of the project and convey that to the other members of the project team – remember that the definition of a project is *a temporary group activity designed to produce a unique product*. It is essential at the start of the project to establish the parameters of the project and to convey this to the rest of the project team. One

consequence of not defining scope is project creep, an outcome that will be discussed later.
5. Lack of definition – may result in the project team being unclear on what has to be achieved.
6. Failure to manage risks – the construction process is one that is subject to a variety of risks, from adverse ground conditions to shortage of materials. These risks must be managed / mitigated in order to achieve a successful conclusion to the project; failure to do this could prove disastrous.
7. Unfamiliar technology – signature building projects with unfamiliar technology will present a greater challenge to the project manager and team than those that use traditional or well-known construction techniques.
8. Inadequate business support – for project success it is essential to have both a robust business case and commitment from the project sponsors.
9. And finally, the position of the project manager is important – not only should the project manager have the responsibility to carry through the project, but also the authority.

1. Initiating the project

The first stage of any project initiation involves putting the resources in place to complete the project successfully and includes (see Table 1.2):

- defining the business model,
- aligning the project with business needs,
- defining outcomes / skills and resources,
- setting objectives, and
- deciding to proceed with the project.

Initiation of the project involves setting the quality and quantity parameters as well as trying to avoid the pit falls that plague many projects. This stage may take place as part of the feasibility study and many months prior to the project moving forward to the next stage.

SWOT *analysis*

SWOT analysis (strengths, weaknesses, opportunities, and threats analysis) is a framework for identifying and analysing the internal and external factors that can have an impact on the viability of a project, product, place or person.

Table 1.2 RIBA Plan of Work 2013 compared with classic project management stages

RIBA Plan of Work 2013	*Classic project management stages*
0 Strategic Definition	1 Initiation
1 Preparation & Brief	1 Initiation
2 Concept Design	2 Planning / organisation
3 Developed Design	2 Planning / organisation
4 Technical Design	2 Planning / organisation
5 Construction	3 Executing / implementation
	4 Monitoring and controlling
6 Handover & Close Out	5 Closing / evaluation
7 In Use	5 Closing / evaluation

The analysis is credited to Albert Humphrey, who tested the approach in 1960s and 1970s at the Stanford Research Institute.

As its name states, a SWOT analysis examines four elements:

1. Strengths – internal attributes and resources that support a successful outcome.
2. Weaknesses – internal attributes resources that work against a successful outcome.
3. Opportunities – external factors the project can capitalise on or use to its advantage.
4. Threats – external factors that could jeopardise the project.

A brainstorming session helps to fill in the SWOT diagram square (Figure 1.8).

PESTLE or PEST analysis

Thought to be a more comprehensive version of SWOT analysis, PESTLE (political, economic, social, technological, legal and environmental) is used as a tool by companies to track the environment in which they're operating or planning to launch a new project (Figure 1.9).

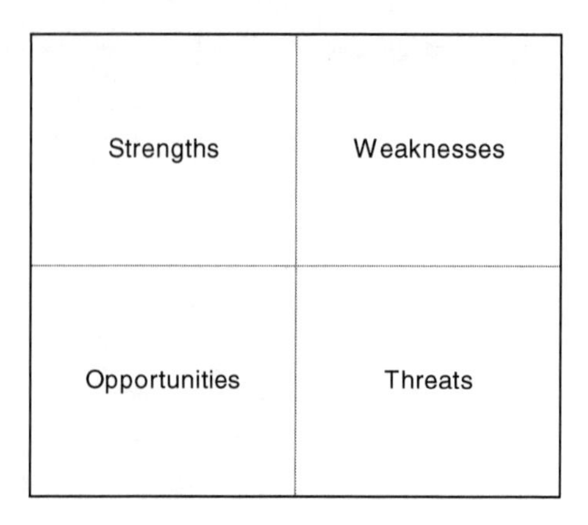

Figure 1.8 SWOT diagram

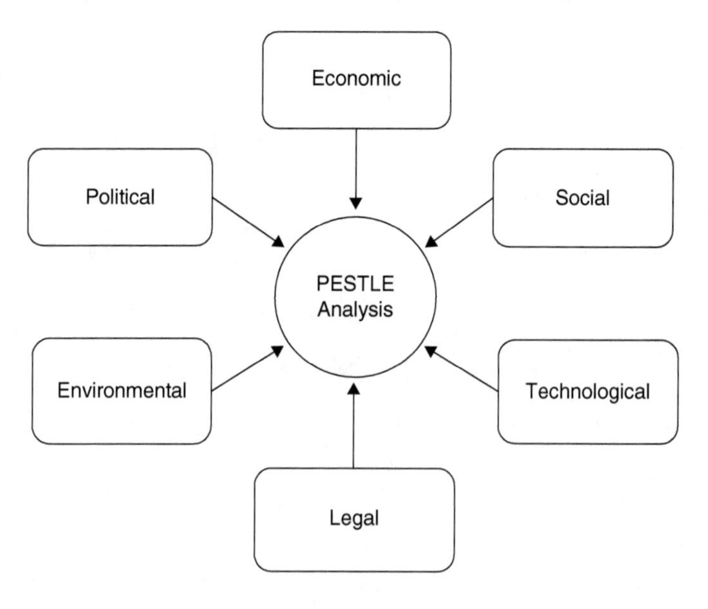

Figure 1.9 PESTLE analysis

2. Planning / organisation

The objective of the planning stage is to evaluate and investigate the best way to achieve the expectations of the client and involves the following tasks:

- organising workload / planning workload / delegation,
- scoping the project,
- drawing up project schedule with key dates,
- defining project objectives,
- defining major deliverables,
- establishing resources,
- carrying out a risk analysis and developing a transparent risk management plan, and
- deciding to proceed with the project.

Developing a project plan

The stages in developing a project plan are:

- Brainstorm a list of tasks to be carried out to complete the project; this can be carried out in conjunction with the project stakeholders.
- Arrange the tasks in approximate order that they will be carried out and convert into an outline plan; give the tasks a reference number or name.
- Estimate, based on previous experiences, the length of time to complete each task and establish task interdependencies.

Table 1.3 shows a sample pre-contract plan.

Project management has been described as 80 per cent planning and the success of this stage often determines the success of the project overall. As previously described earlier in the chapter there are a variety of proprietary software packages and programmes to aid the project manager when planning projects. In addition, the project manager can also utilise:

- Gantt charts / bar charts – these help to work out practical aspects of a project, such as the minimum time it will take to complete a task, and which tasks need to be completed before others can start. In addition it is possible to use them to identify the critical path – the sequence of tasks that must individually be completed on time if the whole project is to deliver on time. Finally, Gantt charts can be used to keep team

Table 1.3 Pre-contract plan

Task	*Description*	*Duration*	*Team members*	*Start*	*Finish*
1.	Select procurement strategy	1 week	DC, AC, CL	23.03.20	27.03.20
2.	Select contract	1 week	DC, AC, CL	23.03.20	27.03.20
3.	Establish tender list	3 days	DC, AC	27.03.20	30.03.20
4.	Preliminary enquiries	1 week	DC	30.03.20	03.04.20
5.	Tender documentation	9 weeks	DC, AC, CL	06.04.20	30.05.20
6.	Tender period	4 weeks	DC	01.06.20	29.06.20
7.	Tender assessment	2 weeks	DC	29.06.20	13.07.20
8.	Contract award		DC, AC, CL	13.07.20	
9.	Project start	4 weeks	DC, AC, CL	10.08.20	07.09.20
Total		22 weeks, 3 days			

members and clients informed of progress. They are simple to update to show schedule changes and their implications, or use to communicate that key tasks have been completed and are easily understood. Figure 1.10 shows a sample Gantt chart.

- PERT analysis – similar in approach to a Gantt chart, a PERT chart is a project management tool used to schedule, organise and co-ordinate tasks within a project. PERT stands for Program Evaluation Review Technique, a methodology developed by the US Navy in the 1950s to manage the Polaris submarine missile program. A similar methodology, the Critical Path Method, was developed for project management in the private sector at about the same time.

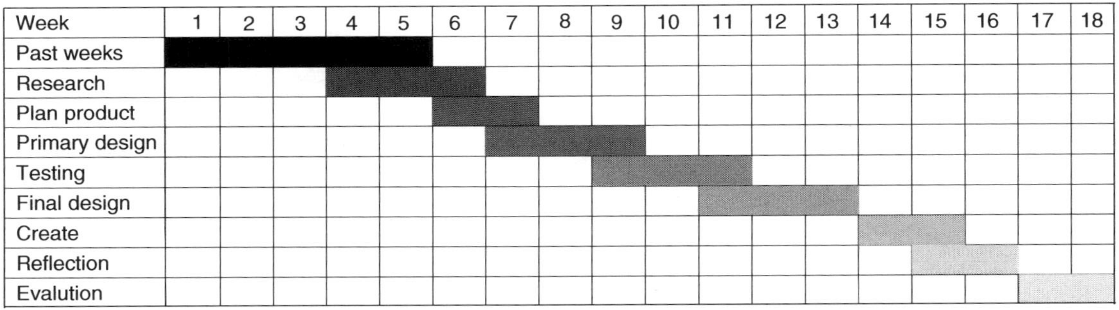

Week	1	2	3	4	5	6	7	8	9	10	11	12	13	14	15	16	17	18
Past weeks																		
Research																		
Plan product																		
Primary design																		
Testing																		
Final design																		
Create																		
Reflection																		
Evalution																		

Figure 1.10 Gantt chart format

3. Executing / implementation

This is when the project gets carried out (built) and involves:

- selecting and appointing the resources to deliver the project with a focus on time / cost / quality and quantity, and
- identifying problems and understanding their impact.

4. Monitoring and controlling

During this phase the metrics are established to compare planned with actual progress of the project, which involves:

- tracking the progress of the project and writing progress reports,
- overseeing project status review sessions,
- compiling contingency plans,
- managing third parties,
- managing change, and
- managing budgets.

5. Closing / evaluation

This process of completion, feedback and review covers:

- signing off the project,
- project review, and
- lessons learned.

THE CONSTRUCTION PROJECT MANAGER

The aim of project management is to ensure that projects are completed at a given cost and within a planned timescale. Before beginning to examine how a construction project manager operates it is first necessary to take a wider look at generic project management skills and techniques.

Project management has many definitions; however, for the purposes of this pocket book it may be regarded as the professional discipline that ensures that the management function of project delivery remains separate from the design / execution functions of a project, and into these generic skills have to be interwoven the specific skills required for construction projects.

Any quantity surveyors reading this book will recognise three of the constraints / objectives discussed previously (Figure 1.1) that need to be controlled by the project manager to deliver project benefits as those generally referred to when selecting an appropriate procurement strategy.

Deciding on the project team structure

To contextualise, the activities that are most commonly involved with construction project management are:

- identifying and developing the client's brief,
- leading and managing project teams,
- identifying and managing project risks,
- establishing communication and management protocols,
- managing the feasibility and strategy stages,
- establishing the project budget and project programme,
- co-ordinating legal and other regulatory consents,
- advising on the selection / appointment of the project team,
- managing the integration and flow of design information,
- managing the preparation of design and construction programmes / schedules and critical path method networks,
- advising on alternative procurement strategies,
- conducting tender evaluation and contractor selection,
- establishing time, cost, quality and function control benchmarks,
- controlling, monitoring and reporting on project progress, and
- administering consultancy and construction contracts.

The role of the project manager and JCT (16)

The role of the project manager is not referred to within JCT (16) nor is there any place within the Articles of the contract to name a project manager; instead the Contract Administrator is referred to as the person with the responsibility of administering, but not necessarily managing, the works. See Figure 1.11.

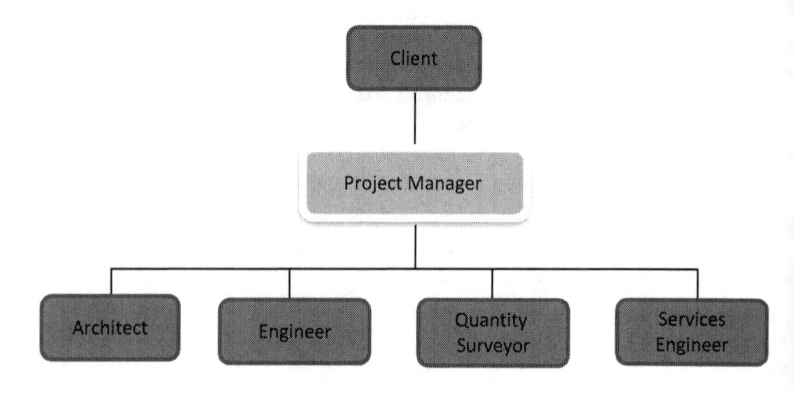

Figure 1.11 Traditional role for project manager in construction project

The role of the project manager and NEC4

The New Engineering Contract (NEC) describes itself as

> *a modern day family of contracts that facilitates the implementation of sound project management principles and practices as well as defining legal relationships. Key to the successful use of NEC is users adopting the desired cultural transition. The main aspect of this transition is moving away from a reactive and hindsight-based decision-making and management approach to one that is foresight based, encouraging a creative environment with pro-active and collaborative relationships.*

The role of the project manager under NEC4 can broadly be defined as to:

- communicate and issue documents in accordance with the contract and generally act in a spirit of trust and co-operation,
- monitor the programme and the contractor's progress against the programme, including reviewing and incorporating proposed alterations,
- fairly manage the compensation event process utilising change management,
- assess payments, and
- manage risk.

NEC4 is intended to provide a modern method for project managers to work collaboratively with employers, members of the design team and contractors.

NEC4 is becoming widely used in both building, and particularly in civil engineering, both in the UK and worldwide. It is generally accepted that the NEC is more than simply a set of clauses; it's more like a management system for building and civil engineering works. What's more, the NEC uses the term project manager to describe the employer's representative who is tasked with the responsibility of administering the works. In addition to the supervision of the works, the NEC Guidance Notes propose that the client / sponsor should appoint a project manager in the early stages of the design sequence to manage the procurement and pre-construction process, not for simply the construction phases, and therefore it follows that the NEC envisages the project manager's role extending from Stage 1 to Stage 6 in the RIBA Plan of Work 2013.

The NEC was launched by the Institution of Civil Engineers in 1993, with the fourth edition in 2017. The boxed set contains a total of twenty-three documents that together make up the new and extended family. It is now the most widely used contract in UK civil engineering and is often used by government departments, such as the Highways Agency, and by local authorities, and has been used on major projects and procurement initiatives. The overall structure is quite unlike that of the JCT and other standard forms, as follows.

Flexibility

- It is intended to be suitable for all the needs of the construction industry.
- It provides for a variety of approaches to risk allocation.
- It is adaptable for some design, full design or no design responsibility, and for all current contract options including target, management and cost reimbursable contracts.

Clarity and simplicity

- The NEC is written in ordinary language, using short sentences with bullet points.
- The simple wording of the documents is deliberately chosen, and lends itself to ready translation into other languages.
- Imprecise terms such as 'fair' and 'reasonable' have been avoided.
- Legal jargon is minimised.
- The actions required from the parties are said to be 'defined precisely', with the aim of avoiding disputes.
- Flow charts are provided to assist usage.

The main changes between NEC3 and NEC4 from the project manager's perspective are concerned with:

- contractor's proposals,
- quality management systems,
- deemed acceptance of programmes,
- payment changes,
- finalising Defined Cost (in cost-based options),
- additional compensation events,
- changes to proposed instructions,
- new secondary options, and
- Schedules of Cost Components (SCC) and Schedules of Short Cost Components (SSCC) changes.

NEC4 places considerable authority in the hands of the project manager, enabling him / her to take action / make decisions on behalf of the client in more than 100 NEC4 clauses that include the following;

- carrying out contract management functions,
- administering the risk register,
- approving all programmes / progress schedules submitted by the contractor,
- monitoring the contractor's progress against the programme,
- approving subcontractors,
- determining the amount due for stage payments,
- evaluating compensation events,
- determining whether acceleration of the works is required,
- dealing with termination of the contract, and
- certifying completion.

If, as the NEC Guidance Notes suggests, a project manager is appointed early in the design sequence, then in addition to the above list, a project manager will be involved in a much wider set of tasks and responsibilities. It is currently thought that instead of blaming the contractor for delays and cost overruns, that in fact higher costs are mainly generated in the early project formulation and pre-construction and therefore early involvement of a project manager when using the NEC4 contract could be the norm.

THE CONSTRUCTION PROJECT MANAGER AND DIGITAL CONSTRUCTION

Digital construction is a new business model that uses digital tools to improve the process of delivering and operating within the built environment and this can include:

- Building Information Modelling (BIM),
- cloud-based software applications,
- robots and drones on-site,
- 3D printing,
- artificial intelligence (AI), and
- blockchain technology.

Overall, construction seems to be behind other sectors when it comes to the application and adoption of digital technologies, with automotive and aerospace leading the way. In a world where projects are continually growing in scale and complexity, digital technologies offer huge potential for improving productivity. For example it is thought that the application of digital technologies could reduce cost and time overruns by 10–15 per cent.

Building Information Modelling (BIM)

It has now been several years since it was announced that Level 2 Building Information Modelling (BIM) was to be mandatory for all central-government-funded projects as part of the Government Construction Strategy 2011. There has been, as is always the case with new government-backed initiatives in UK construction, a lot of debate as to the usefulness of the process. Initially, BIM was poorly sold by the UK government, giving the impression that it was just about expensive systems with case studies emphasising the technology involved with BIM instead of the promotion of greater collaborative working. Consequently the uptake of BIM has not been as rapid or widespread as was initially hoped. The results of the NBS BIM Survey 2019 illustrates that the uptake of BIM overall by the UK construction industry has stalled, with the respondents suggesting the use of BIM on 100 per cent of all projects fell from 18 per cent in 2017/18 to 15 per cent in 2019.

When quizzed further as to the perceived barriers for the adoption of BIM, the following reasons topped the list:

- No client demand – 65 per cent
- Lack of in-house expertise – 63 per cent
- Cost – 51 per cent.

(Adapted from NBS BIM Survey 2019)

There is also anecdotal evidence that the government, at both national and local levels, is not enforcing the 2016 mandate. The phase '*central-government-funded projects*' has come under scrutiny as there is uncertainty as to what projects actually fall into this category. For example, High Speed 2 (HS2) is centrally funded but is not obligated to apply the BIM mandate because it is a non-departmental government agency and therefore subject to a different compliance regime.

To date the project manager has been on the sidelines of BIM development, due in part to the emphasis on the use of BIM in the design process. However, the use of BIM extends beyond initial design issues (see Figure 1.12). As discussed earlier, some of the principal skills of the project manager are collaboration, co-ordination and communication and therefore on the face of it would seem that BIM is potentially a very useful tool for project managers. With BIM as a fundamental enabler for effective integration, it is crucial for project managers to understand how to harness and use it for their projects. The potential for project managers therefore appears to

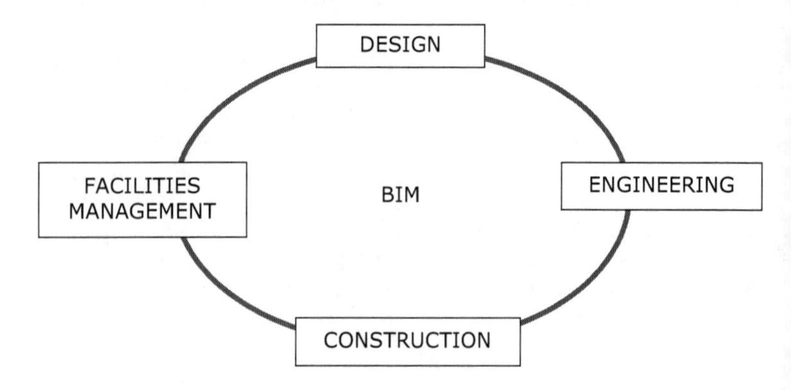

Figure 1.12 The use of BIM extends beyond initial design issues

widen the use of BIM and to ensure that clients are aware of the value added benefits of adopting BIM, both during and after project delivery. The project manager may need to guide a client through the business case for adopting BIM and the required changes to skills, roles and responsibilities. From a skills perspective, BIM is business as usual with the same processes and controls, except for a modified management information system / document protocol, modified roles and responsibilities and a modified procurement strategy. The role of BIM manager should be considered together with the responsibilities they would adopt. The primary issue for project managers is the management, control and interfacing of a data-rich environment, which depending on the maturity level, may all be heavily integrated.

BIM is both a new technology and a new way of working. BIM is a term that has been around for a while in manufacturing and engineering industries, and is now beginning to make an impact in the construction sector. At a strategic level, BIM offers the capacity to address many of the industry's failings including waste reduction, value creation and improved productivity. The early involvement of project managers in the design process enables increased consideration for the constructability and costs as design decisions are being made.

The improved reliability and consistency of BIM-based designs can lower construction costs by enabling many components to be prefabricated off-site in advance. Prefabricated components, by virtue of being made in controlled factory environments, are typically lower cost and lower risk by avoiding unpredictable field conditions (see Chapter 3).

Team members responsible for schedule planning and cost control can use the information in the composite project model throughout the construction process to measure the impacts of design changes and field conditions upon the predicted schedule and budget. Changes to the BIM model can be assessed to generate updated schedules and budget predictions, enabling project managers to better plan and allocate resources in the day-to-day operations at the construction site. In project management, establishing BIM on a project requires a client who understands the upfront costs in return for future benefits, a good BIM protocol and a procurement strategy that constrains silo thinking. The largest single barrier to exploiting BIM is the lack of awareness. Clients are frequently unaware that they can have a major influence on the deliverables from a project. BIM has the potential to impact every aspect of project management.

The process of implementing BIM moves away from using conventional word-processing and computer-aided design (CAD) into the increased use of common standards and product-orientated representations. BIM changes the emphasis by making the model the primary tool for documentation, from which an increasing number of documents, or more accurately reports, such as plans, schedules and bills of quantities, may be derived.

As alluded to earlier, BIM involves more than simply implementing new software; it is a different way of thinking. Successful BIM requires a move away from the traditional communication channels, with all parties, including architects, surveyors and contractors, sharing, and effectively working on, a common information pool. This is a substantial shift from the more traditional convention where parties often work on separate information pools using several different and usually incompatible, software packages. In essence, BIM involves building a digital prototype of the model and simulating it in a digital world. BIM provides a common single and co-ordinated source of structured information, the BIM model, to support all parties involved in the delivery process, whether that be to design, construct, and / or operate. Because all parties involved with a BIM project have access to the same data, the information loss associated with handing a project over from design team to construction team and to building owner / operator is kept to a minimum.

A BIM model contains representations of the actual parts and pieces being used to construct a building along with geometry, spatial relationships, geographic information, quantities and properties of building components, for example manufacturers' details. BIM can be used to demonstrate the entire building life cycle from construction through to facility operation at various levels of detail; 2D, 3D, 4D. Therefore BIM provides a common environment for all information defining a building, facility or asset, together with its common parts and activities. This includes building shape, design and construction time, costs, physical performance, logistics and more. More importantly, the information relates to the intended objects (components) and processes, rather than to the appearance and presentation of documents and drawings.

More traditional 2D or 3D drawings may well be outputs of BIM, however, instead of generating them in the conventional way, i.e. as individual drawings, they could all be produced directly from the model as a 'view' of the required information. BIM changes the traditional process by making the model the primary tool for the whole project team. This ensures that all the designers, contractors and subcontractors maintain their common basis for design, and that the detailed relationships between systems can be explored and fully detailed. Working with BIM will require new skills and these will have to be learned from practice.

BIM is not a silver bullet – it's just as possible to produce a poor model, in terms of its functionality, its constructability or its value, as it is to produce poor drawings, schedules or any other, more traditional, form of information. Also, in the absence of any proactive collaborative management effort, models may end up being prepared to suit the originator as opposed to being structured and presented with all parties to the design and construction team in mind. Ensuring that there is an agreed structure and exchange protocol in place to suit all parties will improve certainty, confidence and consistency. By moving to a shared information model environment, project failures and cost overruns become less likely. BIM certainly means having a better understanding and control of costs and schedules as well as being able to ensure that the right information is available at the right time to reduce requests for information, manage change and limit, or even eliminate, unforeseen costs, delays and claims.

The concept of BIM is not new, having been around since at least the 1980s, when some architects and engineers began to switch from drawing boards to CAD. While this change dramatically increased drawing office productivity, the outputs were still largely paper-based 2D drawings – plans, elevations, sections, isometric views, exploded diagrams, etc. – and still tended to be shared with fellow team members as paper-based documentation.

So, what are the differences between CAD and BIM? CAD is a successful and powerful drafting tool with the following characteristics:

- a drafting tool to produce working drawings,
- a fragmented process containing multiple files,
- a means of creating construction documentation, and
- a system that requires manual co-ordination.

Whereas BIM can be said to be:

- a database of information including details of design, validation, construction and life cycle issues,
- a representation of virtual construction, and
- analytical and quantifiable.

In order to facilitate the introduction of BIM into the UK construction industry, the process has been broken down into levels ranging from 0 to 2 as follows:

- BIM Level 0 – effectively no collaboration or sharing of information with 2D CAD drafting being utilised, mainly for the production of information, with output and distribution being via paper or electronic prints, or a mixture of both. According to the NBS National BIM Report 2016, the majority of the industry is already well ahead of this.

- BIM Level 1 – working at this level is common within the UK construction industry and generally involves the use of a common data environment, such as a collaboration platform, described earlier in this chapter, with which sharing information can be facilitated and managed. Typically this level comprises a mixture of 3D CAD and 2D for drafting production information, with models not being shared between the project team.
- BIM Level 2 – this was the level chosen by the UK government as the minimum target for all work on public-sector work by 2016. The big difference between Level 1 and Level 2 is that Level 2 involves collaborative working, with all parties working and contributing to a shared single model. All parties can access and add to a single BIM model and therefore it is essential that all the parties use a common file format such as COBie (Construction Operations Building Information Exchange). A new BIM Level 2 website was launched in November 2016: http://bim-level2.org/en/.

Table 1.4, taken from the RICS publication Building Information Modelling for Project Managers, illustrates the BIM applications available to the project manager, now and in the future when 4D and 5D modelling become more commonly used.

Clients are often in the best position to lead the introduction of BIM. Understanding the value of building information and its impact on the client's own business is leading many clients to require BIM to specify the standards and methods to be used in its adoption. Clients can also provide clear requirements for facilities management (FM) information to be handed over at project completion more easily with BIM. BIM is equally applicable to support FM and asset management as it is to design and construction. Indeed, the output of the design model may well replace the need for traditional operational and maintenance manuals. Being able to interrogate an intelligent model, as opposed to searching through outdated manuals, perhaps linked to interactive guidance on the repair and / or maintenance process, has obvious advantages.

The principal difference between BIM and 2D CAD is that the latter describes a building by independent 2D views such as plans, sections and elevations. Editing one of these views requires that all other views

Table 1.4 BIM applications available to project managers now and in the future

Stage	*Project manager's BIM role*	*BIM applications*
Briefing, inception and pre-construction	Feasibility analysis (technical and financial)	BIM adoption question, challenges to BIM adoption, concept-stage BIM
	Value engineering	Options selection using BIM, conceptual estimating modelling, energy analysis, design analysis
	Design management	BIM information exchange, 5D (rapid cost feedback to design changes), BIM co-ordination
	Risk analysis and safety	Simulation, virtual reality (VR) and augmented reality (AR)
	Scheduling	4D modelling
	Constructability analysis	4D modelling, virtual mock-ups, VR and AR
	Procurement (design and construction)	BIM skills and capability mapping, BIM enabled supply chain management, constraint analysis
Construction	Phasing and prototyping	4D
	Requests for information (RFIs) and issue resolution	BIM information exchange, BIM co-ordination
	Change management	BIM information exchange
	Monitoring and control	4D and 5D, constraint analysis, progress tracking and production planning
Project closure	Contract and financial closure	Record model
	Project closure	Record model, asset information model
	Handover	Record model, BIM for FM, asset information management

Source: Building Information Modelling for Project Managers (RICS).

must be checked and updated, an error-prone process that is one of the major causes of poor documentation. In addition, data contained in 2D drawings are graphical entities only, such as lines, arcs and circles, in contrast to the intelligent contextual semantic of BIM models, where objects are defined in terms of building elements and systems such as spaces, walls, beams and columns. A BIM model carries all information related to the building, including its physical and functional characteristics and project life cycle information, in a series of 'smart objects'. For example, a lift installation within a BIM would also contain data about its supplier, operation and maintenance procedures. This model can be used to demonstrate the entire building life cycle, and as a result, quantities and shared properties of materials can be readily extracted. Scopes of work can be easily isolated and defined. Systems, assemblies and sequences can be shown in a relative scale with the entire facility or group of facilities. Construction documents such as drawings, procurement details, regulatory processes and other specifications can be easily interrelated.

A building information model can be used by the project manager for the following purposes:

- 3D renderings – these can be easily generated in-house with little additional effort.
- Shop drawings – these can be generated for various building systems, for example, the metal ductwork shop drawings can be quickly produced once the model is complete.
- Building control, planning and fire – these models can be used for review of building projects.
- Clash detection – BIM models are created to scale in 3D space, so all major systems can be visually checked for interferences. This process can verify that piping does not intersect with steel beams, ducts or walls and to graphically illustrate potential failures, leaks, evacuation plans, etc.
- Facilities managers can use BIM for renovations, space planning, and maintenance operations.
- Estimating and quantification – BIM software have built-in cost estimating features. Material quantities are automatically extracted and changed when any changes are made in the model.
- Construction programming – a building information model can be effectively used to create material ordering, fabrication and delivery schedules for all building components.

Project managers should be aware of the potential risks of using BIM, namely:

- Legal risk to determine ownership of the BIM data and how to protect it through copyright and other laws. For example, if the owner is paying for the design, then the owner may feel entitled to own it, but if team members are providing proprietary information for use on the project, their propriety information needs to be protected as well. Thus, there is no simple answer to the question of data ownership; it requires a unique response to every project depending on the participants' needs. The goal is to avoid inhibitions or disincentives that discourage participants from fully realising the model's potential.
- BIM licensing issues can arise. For example, equipment and material vendors offer designs associated with their products for the convenience of the lead designer with the hope of inducing the designer to specify the vendor's equipment. While this practice might be good for business, licensing issues can nevertheless arise if the vendor's design was produced by a designer not licensed in the location of the project (Thompson and Miner, 2007).
- Control of the entry of data into the model and responsibility for any inaccuracies is another fraught area. Taking responsibility for updating BIM data and ensuring its accuracy entails a great deal of risk.
- Requests for complicated indemnities by BIM users and the offer of limited warranties and disclaimers of liability by designers will be essential negotiation points that need to be resolved before BIM technology is utilised.
- The integrated concept of BIM blurs the level of responsibility so much that risk and liability will likely be enhanced. Consider the scenario where the owner of the building files suit over a perceived design error. The architect, engineers and other contributors of the BIM process look to each other in an effort to try to determine who has responsibility for the matter raised. If disagreement ensues, the lead professional will not only be responsible as a matter of law to the claimant, but may have difficulty proving fault with others such as the engineers.
- Professional indemnity insurance and intellectual property rights protection.

Despite the benefits claimed for BIM, there is equally little doubt that the implementation of BIM is a significant sea change for the construction industry and one which does not come with an insignificant expense. This expense is not just limited to the acquisition of new technology programmes, data storage and information security costs, but also training and engagement costs for staff.

Cloud-based software applications

There are a wide range of companies that now offer cloud-based systems. Cloud-based project management solutions manage information and processes on construction and engineering projects and functions include:

- document management,
- access to project drawings,
- bid management,
- issue management,
- handover management,
- control of project correspondence, and
- BIM file management.

Cloud-based solutions allows project members worldwide to create and review documents and other project information from any location via a mobile device.

Robots and drones on-site

Like 3D printing, flying drones can be employed in a number of construction tasks, such as inspection of hard-to-reach areas and work in visually obscured areas. With the price of the technology dropping, drones are being used with increasing frequency. Drones can be used to inspect the condition of refurbishment and structurally unsound projects, performing fly-through surveys of buildings or structures.

3D printing

The industry is still a long way from being able to print an entire building but a number of organisations worldwide, including Massachusetts Institute of Technology, are experimenting with printing building components.

Artificial intelligence (AI)

Artificial intelligence (AI) is a term for describing when a machine mimics human cognitive functions like problem-solving, pattern recognition and learning. The current thinking is that the construction sector is under-digitised compared to other major sectors and that AI could potentially assist in:

- preventing cost overruns,
- risk management,
- big data management, and
- built-asset management.

Blockchain technology

As discussed previously, the construction industry worldwide has a very unenviable reputation for corruption and unethical practice. The management process is complex and with so many players, accountability and transparency is difficult to achieve. Blockchain technology could help to produce more robust processes and supply chain efficiencies together with greater accountability and transparency of transactions. Construction projects rely on various parties to work together to complete a building based on pre-defined specifications, with each party relying on payment based on a quantum meruit basis. A blockchain platform records transactions, agreements and contracts across a peer-to-peer network of computers worldwide. Data is stored as blocks, bound together by coded links into a chain, referred to as a digital ledger. This ledger cannot be altered or amended and anyone in the network can access the latest version at any time. It has been suggested that one of the reason for the failure of Carillion in 2018 was disorganised supply chains and poor subcontractor payment processes. Currently, payment systems such as TraderTransferTrust are being developed that will trigger payment only when work or services have been completed or delivered.

Therefore, the peer-to-peer connectivity of blockchains, combined with smart contract functionality, brings opportunities to streamline project management.

MANAGING MULTIPLE PROJECTS

It is sometimes useful to split large projects into smaller parcels as it reduces complexity and risk. The parcels (phases) could be carried out sequentially, in parallel or overlapping but should be viable pieces of work. They may have dependencies on other parcels of work but it should make sense to manage them separately. The rationale for splitting projects into phases includes:

- mitigating the risk posed by large complex projects,
- the larger project may not be possible as funding may not be available,

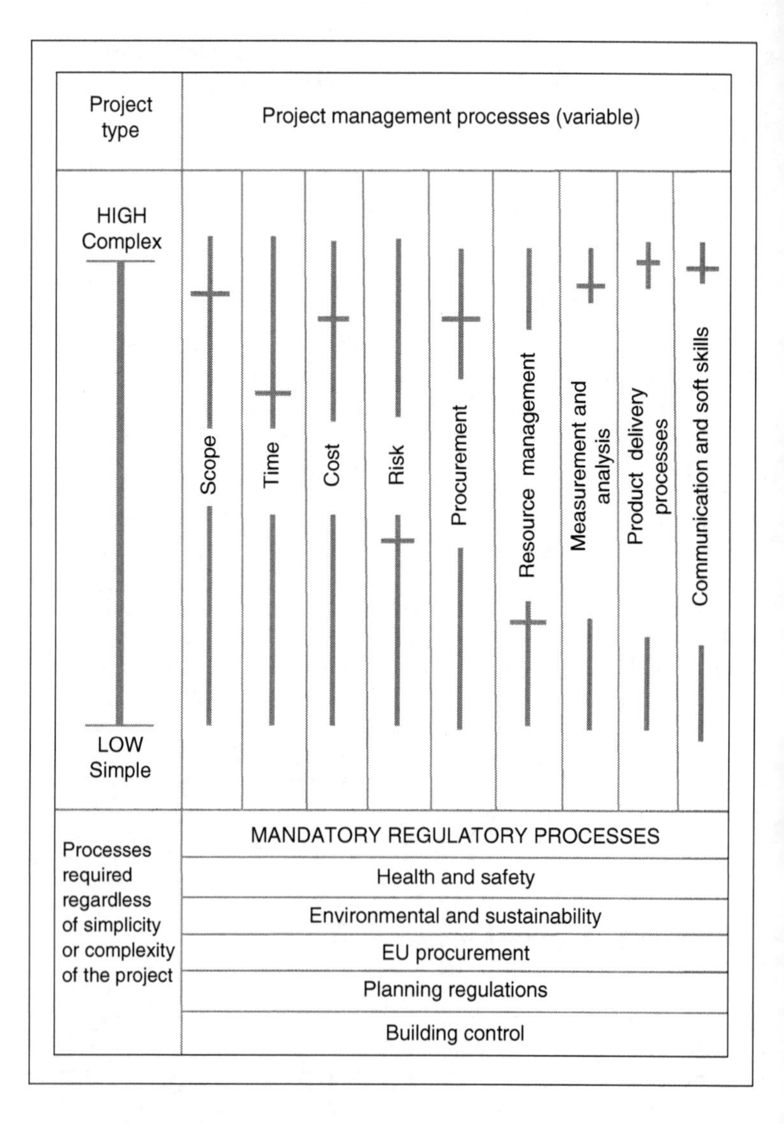

Figure 1.13 Variable and fixed processes

Source: Adapted from BSi PD 6079-4:2006.

- the resources, skilled labour and materials, for example, are unavailable to complete the larger project,
- the exact parameters of the larger project are unclear, and
- the project is due for a phased delivery.

When a larger project is broken down into smaller parcels, these parcels become projects in their own right and the whole package of projects becomes a programme.

As discussed, construction project management comprises a number of processes as illustrated in Figure 1.13. Some of the processes are mandatory regardless of the size or complexity of the project; the extent to which other skills are used will depend on a number of factors related to individual projects and should be applied by the project manager as and when required.

2

Pre-construction / RIBA Plan of Work Stages 0–4 / OGC Gateway Stages 1–3C

The pre-construction phase takes the project from the identification of the business case and the strategic brief, through the feasibility stage and the design development, to include the choice of procurement strategy and the selection of a contractor. It comprises the following:

- Preparation – at this stage the client will be expected to develop and prepare the strategic brief and business case. The project outcomes should be clearly defined and the funding options investigated. In addition the client should also contribute to feasibility studies.
- Design.
- Pre-construction.

EARLY STAGES

There are a number of established approaches / guidance notes to help the project manager through the development process, three of the most popular are compared in Table 2.1. All the guidance notes have been drawn up as a model and road map for the design team and client from the strategic brief to post-practical completion and are freely available on open access.

The CIC Scope of Services is tailored more to allocating roles and responsibilities within a project team rather than a methodology for organising an entire project. On the other hand, the OGC Gateway Reviews are applicable to a wide range of programmes and projects including:

- policy development and implementation,
- organisational change and other change initiatives,
- acquisition programmes and projects,
- property / construction developments,

Table 2.1 Approaches to development process compared

RIBA Plan of Work 2013	*OGC Gateway*	*CIC Scope of Services 2007*
0 Strategic Definition	Gate 0 Strategic assessment Gate 1 Business justification	Stage 1 Preparation
1 Preparation & Brief	Gate 2 Procurement strategy	Stage 1 Preparation
2 Concept Design		Stage 2 Concept
3 Developed Design		Stage 3 Design development
4 Technical Design	Gate 3 Investment decision	Stage 4 Production information Stage 5 Manufacture, installation and construction information
5 Construction		
6 Handover & Close Out	Gate 4 Readiness for service	
7 In Use	Gate 5 Benefits evaluation	Stage 6 Post practical completion

- IT-enabled business change, and
- procurement using or establishing framework arrangements.

The OGC was subsumed into the Cabinet Office as part of a round of government spending cuts in May 2010 and the OGC Gateway Review was archived and will not be updated. The review uses a series of interviews, documentation reviews and teams' experience to provide valuable additional perspectives on the issues facing the project team, and an external challenge to the robustness of plans and processes. Despite the above, the Gateway Review is still referred to in a number of government policy documents and will therefore be referred to also within this pocket book.

The Stages of the RIBA Plan of Work (2013) will be used throughout the remaining chapters and it should be noted that the 2013 RIBA Plan of Work has the ability to carry out limited customisation of certain tasks of the plan, namely

- Core Objectives,
- Procurement,*
- Programme,*
- (Town) Planning,*
- Suggested Key Support Tasks,
- Sustainability Checkpoints,
- Information Exchanges, and
- UK Government Information Changes.

* Customisation of tasks possible.

With the introduction of the customisable RIBA Plan of Work, some task bars are fixed to provide a degree of continuity from one plan to the next, whereas other task bars can be selected from pull-down menus or switched on or off. Pull-down menus are provided for the Procurement, Programme and Planning task bars. Selecting a procurement route inserts specific procurement activities into the customised plan. In the initial version, the online tool selects the programme task bar based on the procurement route highlighting the stages likely to overlap or be undertaken concurrently. The Planning task bar recommends that planning applications are made at the end of Stage 3 but also allows applications to be made at the end of Stage 2.

The Sustainability and UK Government Information Exchange task bars can be switched on or off depending on the project and it is also anticipated that most customised Plan of Works will utilise the Sustainability bar. Customisation can be performed by downloading the Plan of Work from: www.ribaplanofwork.com/CreatePlan.aspx

RIBA PLAN OF WORK (2013)

The RIBA Plan of Work (2013) is the latest in a long line of documents which over the years has become the industry standard. The plan is divided into eight stages and for each stage the tasks that should be completed are listed in the form of task bars. The stages are:

Task bar

0 Strategic Definition,
1 Preparation and Brief,
2 Concept Design,
3 Developed Design,
4 Technical Design,
5 Construction,
6 Handover and Close Out, and
7 In Use.

Task bar 8: BIM Information Exchanges

The UK Government Information Exchanges Task Bar 8 has been introduced into the plan of work to encourage consideration of by the project team of the stages that the UK Government requires information to be exchanged. This task bar highlights the UK Government's particular views on this subject as outlined in its 2011 Construction Strategy and subsequent publications. The UK Government recognises that, as a client, it does not need to be involved in every information exchange but it requires particular and specific information at certain stages in order to answer the questions pertinent to a given stage. Furthermore, the UK Government is seeking data-rich information that can be used post-occupancy to manage its entire estate and to allow stringent benchmarking activities to occur. This task bar is selectable and can be switched on or off as required.

CHOICE OF PROJECT TEAM MEMBERS

The project manager will know the skill set that ideally they would like as part of the project team but often the choice of team members may involve compromise. The project manager should therefore be prepared to consider and do the following;

- adapt the structure of the project team to match the availability of resources,
- realise that fewer of the right people are better than more of less suitable individuals,
- determine whether team members are:
    - available for the duration of the project,
    - able to communicate effectively,
    - available full-time or will be involved in several project simultaneously, and
    - able to work effectively as part of a team.

ROLES OF PROJECT TEAM MEMBERS

The roles of the various project team members will vary according the procurement strategy and the form of contract. The project manager should be aware of the following roles and responsibilities.

Client or project sponsor

Clients come in all shapes and sizes and the extent to which they will expect to engage with the development process also varies widely. However, as a general rule a client will be expected to:

- articulate the project vision and communicate to different team members,
- define overall aims and objectives of the project,
- set up the selection process for any external independent client advisor(s), and help in their selection,
- co-ordinate the in-house and client advisor input to the assessment of need and options, business case and budget,
- present information to the board (or chief executive),
- lead in preparation of project brief,
- set up structures for managing the in-house and project teams,
- identify all users and stakeholders, and ensure they are involved and consulted – particularly important when carrying out value engineering / management workshops,
- ensure decision-makers understand their responsibilities and have enough time, resources and information,
- confirm the project is needed and then commit to build, and
- start planning for occupation, especially if organisational change is anticipated.

See also Table 2.2.

Project manager

The activities most commonly involved with construction project management typically include:

- identifying and developing the client brief,
- leading and managing project teams,
- identifying and managing project risks,

Table 2.2 Client's role in pre- and post-contract stages

Preparation	*Design*	*Pre-construction*	*Construction*	*In Use*
Strategic definition & brief	Design / concept / design development	Technical design	Construction	Handover / In Use
Develop business case for project Appoint adviser Define client's responsibilities Project brief Appointment of project manager (if appropriate) Appointment of design and cost consultants Procurement strategy Value management Resources	Procurement strategy Design overview Cost control overview Whole life costs Value engineering Time control overview Quality control overview Appointment of constructors Confirming the business case	Design overview Cost control overview Time control overview Quality control overview Change control overview	Design overview Cost control overview Time control overview Quality control overview	Commissioning Occupation and takeover

Source: Adapted from RICS (2013), Information Paper, *The construction sectors and roles for the chartered quantity surveyor*, RICS Publishing.

- establishing communication and management protocols,
- managing the feasibility and strategy stages,
- establishing the project budget and project programme,
- co-ordinating legal and other regulatory consents,

- advising on the selection / appointment of the project team,
- managing the integration and flow of design information,
- managing the preparation of design and construction programmes / schedules and critical path method networks,
- advising on alternative procurement strategies,
- conducting tender evaluation and contractor selection,
- establishing time, cost, quality and function control benchmarks,
- controlling, monitoring and reporting on project progress, and
- administering consultancy and construction contracts.

The above list is by no means exhaustive.

Architect / lead designer

Traditionally in the UK the architect has been regarded as the leader of the design team and the first person to be appointed by the client at the start of a new project. So much so, that traditional single stage tendering is sometimes referred to still as 'architect-led tendering'.

Perhaps the most difficult part of the architect's role is to interpret a client's user requirements and transform them into a building. Unlike the rest of Europe most architects work within private practice, with few working for contractors or developers, the UK is home to some of the largest firms of commercial architects in the world. The Royal Institute of British Architects and the Royal Incorporation of Architects in Scotland are the professional bodies for architects in the UK. Architects can also act as contract administrators although increasingly this role is being taken over by others.

Contract administrator

Contract administrators make sure the parties employ due diligence to comply with the terms, conditions, rights and obligations of the contract. They also co-ordinate any changes to the agreement that might occur over the course of the contract and perform the closeout process when both parties have met their obligations.

Contract compliance is a large part of the administrator's job. They make sure that all performance obligations specified in the contract are being met. Monitoring and tracking performance over the course of the contract is usually accomplished through the use of electronic document management systems and spreadsheets.

As part of the monitoring process, the contract administrator inspects goods when delivered to make sure the delivery is per the agreement. Depending on the type of contract, he or she might inspect services rendered or visit the job site to ensure that work is being done according to the contract agreement. Payment usually is not made for goods or services until the administrator has determined that the terms of the agreement have been met.

Information manager

The information manager is responsible for the establishment of the Common Data Environment (CDE) which is used to exchange all project information, not just Building Information Models. Their role is to:

- establish a CDE including processes and procedures to enable reliable information exchange between project team members, the employer and other parties,
- establish, agree and implement the information structure and maintenance standards for the information model,
- receive information into the information model in compliance with agreed processes and procedures,
- validate compliance with information requirements and advise on non-compliance,
- maintain the information model to meet integrity and security standards in compliance with the employer's information requirement, and
- manage CDE processes and procedures, validate compliance with them and advise on non-compliance.

BIM manager / co-ordinator

With a wider set of responsibilities than the information manager, being more closely aligned to design, the BIM manager's primary function is to manage the process of virtually constructing a building and accurately documenting the design contract documents. This encompasses managing a team of production professionals, designers and technicians of multiple disciplines and owning the construction documents set through as-built submittals. It is also critical for them to lead model management and BIM planning, collaboration and co-ordination on projects they are leading. The position becomes the go-to person on the project for modelling, documentation and verifying design.

The BIM manager's responsibilities include:

- design team compliance with the project's BIM standards,
- manage delivery of the appropriate visuals to team members to support their work,
- create and capture evidence of BIM values which influence the commercial outcomes on the project and extracting data from the model to contribute directly to support monthly reporting,
- interrogate the design input of the model to identify clashes and produce and manage a clash register,
- oversee the extraction of key data from the project model to produce scheduled material quantities / take-offs,
- interface with and support the procurement team,
- interface with the project planning software to create a virtual build of the 3D model,
- connect BIM to the on-site activities through the site management team,
- optimise the site logistics through the model when planning temporary works, and
- build a data set during the design and construction to reflect the needs of the client's asset manager.

The RICS has developed a BIM Manager Certification in response to industry requirements to have a kite-mark that demonstrates the skills and competence of construction professionals in using BIM. A similar programme is also available at the Building Research Establishment (BRE).

Cost consultant / quantity surveyor

It is not uncommon for practices, particularly the larger quantity surveying ones, to supply both quantity surveying and project management services for the same project / client. The quantity surveyor is responsible for the following:

- Strategic definition / preparation / brief:
    - liaise with clients and the professional team,
    - advise on cost, and
    - prepare initial budget / cost plan / cash flow forecasts.
- Concept / developed design:
    - prepare and maintain cost plan, and
    - advise design team on impact of design development on cost.

- Technical design:
  - liaise with professional team,
  - advise on procurement strategy,
  - liaise with client's legal advisors on contract matters,
  - prepare tender documents,
  - define prospective tenderers,
  - obtain tenders / check tenders / prepare recommendation for client, and
  - maintain and develop cost plan.
- Construction:
  - visit the site,
  - prepare interim valuations,
  - advise on the cost of variations,
  - agree the cost of claims, and
  - advise on contractual matters.
- Handover and in use:
  - arrange release of retention funds,
  - prepare the final account, and
  - prepare recommendations for liquidated and ascertained damages.
- Supplementary services:
  - preparation of mechanical and electrical tender documentation,
  - preparation of cost analyses,
  - advise on insurance claims,
  - facilitate value management exercises,
  - prepare life cycle calculations / sustainability,
  - capital allowance advice / VAT, and
  - attend adjudication / mediation proceedings.

Structural engineer

A structural engineer is involved in the design and supervision of the construction of all kinds of structures such as houses, theatres, sports stadia, hospitals, bridges, oil rigs, space satellites and office blocks. The specialist skills of a structural engineer will include calculating loads and stresses, investigating the strength of foundations and analysing the behaviour of beams and columns in steel, concrete or other materials to ensure the structure has the strength required to perform its function safely, economically and with a shape and appearance that is visually satisfying.

Civil engineer

Civil engineers are involved with the design, development and construction of a huge range of projects in the built and natural environment. Consulting civil engineers liaise with clients to plan, manage, design and supervise the construction of projects. They can run projects as project manager. Within civil engineering, consulting engineers are the designers; contracting engineers turn their plans into reality. Consulting civil engineers provide a wide range of services to clients. Typical work activities include:

- undertaking technical and feasibility studies and site investigations,
- developing detailed designs,
- assessing the potential risks of specific projects, as well as undertaking risk management in specialist roles,
- supervising tendering procedures and putting together proposals,
- managing, supervising and visiting contractors on-site and advising on civil engineering issues,
- managing budgets and other project resources,
- managing change, as the client may change their mind about the design, and identifying, formalising and notifying relevant parties of changes in the project,
- scheduling material and equipment purchases and delivery,
- attending public meetings and displays to discuss projects, especially in a senior role,
- adopting all relevant requirements around issues such as building permits, environmental regulations, sanitary design, good manufacturing practices and safety on all work assignments,
- ensuring that a project runs smoothly and that the structure is completed on time and within budget, and
- correcting any project deficiencies that affect production, quality and safety requirements prior to final evaluation and project reviews.

In many countries, civil engineers are subject to licensure, and often, persons not licensed may not call themselves civil engineers.

Construction manager

Depending on the chosen procurement strategy (see the section on risk and procurement strategies later in this chapter) the client may elect to use a construction manager. They will be employed directly by the client and will be responsible for:

- selection of the consultants,
- determining the number and type of work packages,
- management of the procurement process,
- site management / organisation / programming, and
- supervision of the works packages on-site.

Main contractor / subcontractors / suppliers

The responsibilities / obligations of the main contractor are set out in the various standard forms of contract that are used in the construction industry, for example the JCT (16) suite, NEC4, ACA, etc., and will typically include:

- carrying out the works,
- control of the works,
- agreeing payments,
- carrying out variations, and
- agreeing the final account.

Currently, most work is carried out by subcontractors working under the direction of the main contractor although a small number of standard forms of contract still have provision for the appointment of nominated subcontractors and suppliers.

Other parties / organisations with whom the project manager may expect to engage with include the principal designer and principal contractor in the context of the CDM Regulations (2015) (see later in this chapter).

Environmental health officer

An environmental health officer (EHO) aims to make sure that people's living and working surroundings are safe and hygienic. EHOs deal with a wide range of construction related issues including:

- environmental protection and pollution control,
- noise control,
- health and safety at work,
- public health,
- waste management,
- housing standards,

- investigating accidents at work, and
- working closely with housing, building control, trading standards and waste management officers and the Health and Safety Executive.

Building control / warrant officer

The main function of building control is to ensure that the requirements of the building regulations are met in all types of non-exempt development by way of examining the drawings, specifications, etc., in addition to checking the work at various stages as it proceeds. Most building control officers are now actively involved at design stage for many schemes and are acknowledged to provide valuable input at all stages of development.

Local planning authorities (LPAs)

Local planning authorities (LPAs) are responsible for dealing with planning applications in accordance with the relevant legislation and structure plan in their region.

Fire safety inspector

The fire safety inspector will carry out all categories of fire protection audits and assessments of plans and premises as directed in accordance with the relevant legislation.

The police

The police may have to be notified if it is likely that the project works will cause disruption to traffic or pedestrians.

PREPARATION

Strategic Definition / Preparation (RIBA Stages 0 & 1)

Strategic Definition is a new stage in the RIBA Plan of Work (2013), although some of the tasks have been taken from the previous version of the plan, in which a project is strategically appraised and defined before a detailed brief is created. This is particularly relevant in the context of sustainability, when dealing with refurbishment or extensions. The items that have to be addressed are:

- identify business case,
- initial thoughts about project team,
- establish project programme, and
- sustainability check points – developed from the Sustainability Checkpoints included in the 2011 *Green Overlay to the RIBA Outline Plan of Work 2007*; clarifies activities required to achieve the sustainability aspirations such as reducing the carbon emissions to the building.

Terms of engagement / appointment

Because of the variety and range of services that fall within the compass of the project manager, it is in the interest of both the client and the project manager to clearly define the terms of engagement and scope of services at the earliest possible stage.

The documents listed below are some of the available standard forms of appointment for project managers that attempt to define the scope and nature of project management services and typically provide the option for basic and additional services:

- RICS Project Management Agreement (3rd Ed) (1999),
- APM Terms of Appointment for a Project Manager (1998),
- NEC Professional Services Contract (2013), and
- RIBA Form of Appointment for Project Managers (2004).

An important part of the written appointment of the project manager is the schedule of duties or scope of services to be provided, as these can vary widely. For all project manager appointments, the client and the project manager should review the range of services that are available and agree upon a definitive scope of services as early as possible.

The use of standard schedules of services is useful, especially where it is co-ordinated with the other roles on the project. This aids clarity for each of the participants, which improves their understanding of their role, and the roles of others, as well as providing a visible demonstration of the interdependencies and interrelationships between tasks which, in turn, should lead to a reduction in risk.

There should be a clear recognition of the stages and gates required by the client or proposed by the project manager (RIBA Plan of Work and OGC Gateway), and the levels of authority vested in the project manager.

Pricing services for well-defined and straightforward projects is relatively simple, in which case either a lump sum or percentage fee might be appropriate. When using a percentage it is important to define and agree what the percentage will be applied to, e.g. the approved budget, the accepted tender for the construction contract or the final cost of the works. As the complexity of the project increases, other methods of reimbursement may be more reliable. If a construction professional takes an appointment as a project manager they will need to ensure that appropriate professional indemnity insurance is in place, as required by the regulations of their professional body.

As a guide, the RICS Guidance Note *Appointing a Project Manager* recommends the following key elements are incorporated into any form of appointment:

- the parties to the appointment,
- applicable law,
- the services to be performed (which may be split between basic services and additional services),
- general obligations (including standards to be exercised),
- provision for instructions and changes,
- health and safety, statutory requirements and prohibited materials (if applicable),
- design responsibility (if applicable),
- limitation on liability,
- collateral warranties and rights of third parties,
- key (and other) personnel,
- client obligations,
- payment (amounts and periods),
- authority levels,
- insurances,
- copyright and confidentiality,
- assignment, transfer of rights and obligations,
- subcontracting,
- suspension and termination,
- dispute resolution, and
- notices.

BSRIA Soft Landings and Government Soft Landings (GSL)

Handover is the crucial stage of any building project, as it offers the opportunity to fine-tune and debug the building's systems and explain to facilities managers and occupants how to get the best out of their new environment.

It can be especially valuable in buildings where, for example, low-energy, passive or mixed-mode ventilation systems are used as these can involve greater occupant interaction and fine-tuning than a fully air-conditioned office. In practice it means having members of the original team on-site for a period – typically up to two days a week for the first two months – as the occupants move in. Crucially, the project manager should be aware this transition needs to be considered throughout the development of a project, not just at the point of handover and the client should commit to adopting a Soft Landings or GSL strategy (see Table 2.3) in the very early stages in order that an appropriate budget can be allocated and appointment agreements and briefing documents can include relevant requirements. See also Chapter 4.

Professional indemnity insurance for project managers

Construction project managers, by the nature of their work, require special consideration when it comes to professional indemnity insurance. They are often appointed by the client, who is often, but not always, the developer, but equally project managers are employed by government / local authority departments, housing associations or are self-employed. Where the project manager is a direct employee of a private sector construction or development company then the professional indemnity insurance taken out by that firm should automatically cover these activities. However, if the firm for which they are working is a non-regulated company they should ensure that:

1. the company has professional indemnity insurance, and
2. the project management activities have been fully declared to insurers and are indemnified.

The danger comes for project managers who are employed by public-sector employers, as many local authorities, housing associations, governmental departments and the like do not carry professional indemnity insurance on the grounds that their work is internal and you cannot claim against yourself. The project manager, however, comes into contact with a vast number of external organisations and people, any and all of whom could make a claim against them if things go wrong. Because of the interaction of project manager and organisations / persons outside of the organisation for which they work, it is essential that professional indemnity insurance is arranged, even where the project manager is a public sector employee where such cover is not normal.

Table 2.3 Statutory approvals / planning permission

RIBA Plan of work 2013	*Soft Landings*	*Soft Landings supporting activities*
0 Strategic Definition		
1 Preparation & Brief	Identify all actions needed to support procurement.	Define roles and responsibilities. Explain Soft Landings to all participants, identify processes.
2 Concept Design	Design development to support the design as it evolves.	Review past experience. Agree performance metrics. Agree design targets.
3 Development Design	Scheme design reality check.	Review design targets. Review usability and manageability.
4 Technical Design	Technical design reality check.	Review against design targets. Involve the future building managers. Include additional requirements related to Soft Landing procedures Include evaluation of tender responses to Soft Landing requirements
5 Construction		Confirm roles and responsibilities of all parties in relation to Soft Landing requirements.
6 Handover & Close Out	Pre-handover reality check. Pre-handover – prepare for building readiness; provide technical guidance. Post-handover review.	Include FM staff and / or contractors in reviews. Demonstrate control interfaces. Liaise with move-in plans.

Table 2.3 (Cont.)

7 In Use	Aftercare in the initial period – support in the first few weeks of occupation. Years 1–3 Aftercare – monitoring review, fine-tuning and feedback.	Incorporate Soft Landing requirements. Set up home for resident on-site attendance. Operate review processes. Organise independent post-occupancy evaluations.

Adapted from UBT / BSRIA The Soft Landings Framework (2014)

PREPARATION / BRIEFING

The project manager should make the client fully aware of the importance of the briefing process in achieving a finished project that will match their expectations. The RIBA Plan of Work Stage 0 recommends that the core objectives of this stage are to:

- identify the client's business case,
- identify the client's strategic brief, and
- identify other core project requirements, for example sustainability targets.

The business case

Clients / project sponsors come in a wide range of types and with varying degrees of experience when it comes to construction / development. Whereas experienced clients may have their own in-house team to prepare a business case, others, who are less experienced, may need assistance to do this.

The strategic brief

The business case is not just about establishing aims and objectives and should address the following points:

- Is there a need for the project and is there a clear understanding of the outcomes or deliverables? Deliverables are the tangible outputs of the project; they are what the project will physically deliver and define the boundaries of the project. As such, the client needs to understand the size of the undertaking. The boundaries should be clearly defined to prevent project creep and misunderstandings later. Lack of clarity about the scope will lead to misunderstandings further into the project and pressure to deliver things the project was not designed to do. (Outcomes of course come in a variety of forms, for example increased production space, increased output / productivity / sales, increased market profile, personal prestige.)
- Define the benefits of the project. It's the use to which the deliverables are put which leads to benefits. The benefits of the project need to be clearly articulated and linked to the client's or sponsor's objectives.
- What resources will be needed to deliver the project? Money is usually the headline resource; in addition people and equipment should be considered. At this point initial consideration can be given to assembling the project team.
- Identify any constraints. How long will the project take? It would be useful to produce a high-level Gantt chart showing the key milestones. When are important deliverables and benefits expected to happen? When is the project expected to close out? Any anticipated planning or regulatory issues could be considered.

Factors that may influence the preparation of the strategic business case include:

- the general economic climate, including interest rates, short- and medium-term forecasts,
- government intervention, for example planning issues, release of green belt land for development,
- demographics and changing needs of the markets, e.g. the rise of Internet C2B and B2B business,
- new entries to the market / increased competition, and
- issues surrounding the availability and cost of materials and labour.

Feasibility studies / business case development

Following the agreement of the strategic business case, the project manager can now start to assess the options for meeting the stated deliverables with the preparation of the feasibility study or report. This stage / report is one of the key milestones of a project and should include the following:

- order of capital cost estimate (NRM 1),
- order of repairs and maintenance costs estimate (NRM 3),
- schedule of accommodation,
- value engineering / value management recommendations,
- risk assessments (note that NRM suite discourages the use of the term 'contingencies' in favour of a more carefully considered risk assessment),
- sensitivity analyses,
- site-related matters including potential site assessments if applicable, geo-technical surveys,
- health and safety study,
- exclusions,
- Environmental Impact Assignment (if appropriate),
- EU Public Procurement requirement (if appropriate), and
- potential for use of IT packages, including BIM.

IT / software packages / BIM

Previously, in Chapter 1, a number of IT and IT-based packages, such as Newforma and Microsoft Project, were described as being available to aid the project manager. During the past five years or so, however, the project management software market has been dominated by Building Information Modelling (BIM). BIM, which has been around for a number of years, was given added prominence when the UK government announced that from 2016 all government construction procurement must be BIM enabled. The project manager should understand that clients are often in the best position to lead the introduction of BIM. Understanding the value of building information and its impact on the client's own business is leading many clients to require BIM to specify the standards and methods to be used in its adoption. Clients can also provide clear requirements for facilities management information to be handed over at project completion more easily with BIM. At a strategic level, BIM offers the capacity to address many of the industry's failings including waste reduction, value creation and improved productivity. BIM changes the emphasis by making the model the primary tool for documentation, from which an increasing number of documents, or more accurately 'reports', such as plans, schedules and bills of quantities, may be derived.

One of the major challenges for a traditionally confrontational construction industry is that BIM requires a move away from the traditional workflow, with all parties (including architects, surveyors and contractors) sharing, and effectively working on, a common information pool. This is a major shift from the more traditional convention where parties often work on separate information silos using several different (and usually incompatible) software packages.

In its purest form, BIM provides a common single and co-ordinated source of structured information to support all parties involved in the delivery process, whether that be to design, construct and / or operate. Because all parties involved with a BIM project have access to the same data, the information losses associated with handing a project over from design team to construction team and to a building owner / operator is kept to a minimum. A BIM model contains representations of the actual elements and components being used to construct a building along with geometry, spatial relationships, geographic information, quantities and properties of building components (for example manufacturers' details). BIM can be used to demonstrate the entire building life cycle from construction through to facility operation.

BIM provides a common environment for all information defining a building, facility or asset, together with its common parts and activities. This includes building shape, design and construction time, costs, physical performance, logistics and more. More importantly, the information relates to the intended objects (components) and processes, rather than relating to the appearance and presentation of documents and drawings, although traditional 2D or 3D drawings may well be outputs of BIM.

BIM changes the traditional process by making the model the primary tool for the whole project team. This ensures that all the designers, contractors and subcontractors maintain their common basis for design, and that the detailed relationships between systems can be explored and fully detailed.

By moving to a shared information model environment, project failures and cost overruns become less likely. BIM certainly means having a better understanding and control of costs and schedules as well as being able to ensure that the right information is available at the right time to reduce requests for information, manage change and limit (or even eliminate) unforeseen costs, delays and claims.

BIM is equally applicable to support facilities management and asset management as it is to support design and construction. Indeed, the output of the design model may well replace the need for traditional operation and maintenance manuals. Being able to interrogate an intelligent model, as opposed to searching through outdated manuals, perhaps linked to interactive guidance on the repair and / or maintenance process has obvious advantages.

What skills do project managers need to implement BIM? Establishing BIM on a project requires a client who understands the upfront costs (in return for future benefits), a good BIM protocol and a procurement strategy that constrains silo thinking. The project managers may need to guide a client through the business case for adopting BIM and the changes to skills, roles and responsibilities. From a skills perspective, BIM is business as usual with the same processes and controls, except for a modified management information system / document protocol, modified roles and responsibilities, and a modified procurement strategy. The role of BIM manager should be considered together with the responsibilities they would adopt. The primary issue for project managers is the management, control and interfacing of a data-rich environment that, depending on the maturity level, may all be heavily integrated.

What are the benefits for the project manager? These include:

- Updates can be dynamic, removing some risks associated with data management.
- Increased confidence and risk reduction, such as design co-ordination (e.g. structure and services), construction logistics and timelines.
- Cost and programme implications, ideally, would be real time (but will need a sense check to understand all the implications, e.g. whether weekend working is required).
- Improved communications between the project manager, stakeholders, owners, end users, third parties and within the project team.
- The project team and client can visualise, simulate and analyse a project before actual construction begins. It allows the visualisation of phasing and subsequent impact on logistics, cash flow and sales (e.g. you cannot sell prime residential apartments if they face construction works).
- Integration of design and programme increases confidence in completion dates and refines project preliminaries.
- Depending on a project's position on the BIM maturity model, change management should be simplified and easily identifiable (what will not be evident is why the change is being considered), and the change impact will immediately be reflected throughout the model.
- If performing design management, BIM will co-ordinate a change made anywhere in the model – in 3D views and drawing sheets, schedules and elevations, sections and plans – and scope gaps can be checked for.

- For performance management of the design and construction team, the design updates are readily available (or as parts are designed off-line, tested and uploaded to BIM) for performance review and checking against programme.
- BIM is updated during construction to create an 'as-built' record and the model becomes a record to support facilities management. The objects link to data about each component, which facilitates delivery of the building record documents.

Sustainability / green issues

Project teams have to negotiate a complex set of environmentally driven legislation requirements when deciding the most cost-effective approach to schemes. As a minimum all projects must comply with Part L of the Building Regulations. Additionally the government sees planning systems as an effective way of improving the environmental performance of projects over and above Part L, helping to meet its carbon reduction targets.

Local authorities can demand a variety of measures to improve building environmental performance as a condition of planning. This includes meeting a minimum percentage, typically 10 per cent, of building energy needs from on-site renewables. Other local authorities require buildings to better Part L requirements by a minimum percentage, which can vary from 10 per cent up to 25 per cent in London. Some local authorities stipulate a minimum BREEAM rating as a condition of planning which is also a condition for funding in many public sector schemes.

The project manager should be aware of the need to comply with sustainability legislation and issues. This may be necessary for a number of reasons including:

- LPAs require that new / proposed developments obtain stipulated BREEAM ratings in order to be granted planning permission.
- To comply with corporate statements on sustainability / ethical matters.
- Demand from clients.

Increasingly clients as well as end users are requesting improved sustainability performance from their buildings, over and above the regulatory requirements arising from changes in the Building Regulations. Methodologies such as BREEAM (Building Research Establishment Assessment Method) and LEED (Leadership in Energy and Environmental Design) are often used as the vehicle for achieving these improvements. However, these tools are largely environmentally biased, and it is important that the wider social and economic dimension of sustainability is also considered. It is strongly recommended that these issues are considered holistically by the project manager at an early stage in project inception and taken forward in an integrated manner. From a sustainability perspective, refurbishment projects are increasingly expected to achieve design standards of new build projects including:

- improved quality and value for money,
- reduced environmental impact and improved sustainability,
- healthy, comfortable and safe internal and external environments that offer high occupant satisfaction and productivity,
- low costs-in-use, and
- a flexible and future-proofed design.

The measures adopted to assess sustainability performance – and developers and design teams are encouraged to consider these issues at the earliest possible opportunity – are:

- BREEAM (Building Research Establishment Assessment Method),
- BREEAM-In-Use – for existing buildings,
- LEED (Leadership in Energy and Environmental Design), and
- EcoHomes points / Code for Sustainable Homes.

BREEAM

BREEAM has been developed to assess the environmental performance of both new and existing buildings. BREEAM assesses the performance of buildings in the following areas:

- management – overall management policy, commissioning and procedural issues,
- energy use,
- health and wellbeing,
- pollution,
- transport,
- land use,
- ecology,
- materials, and
- water, consumption and efficiency.

In addition, unlike EcoHomes points, BREEAM covers a range of building types such as;

- offices,
- industrial units,
- retail units, and
- schools.

Other building types such as leisure centres can be assessed on ad hoc basis.
In the case of an office development the assessment would take place at the following stages:

- design and procurement,
- management and operation,
- post-construction reviews, and
- building performance assessments.

A BREEAM rating assessment comes at a price and according to the BRE website, the fee scale for BREEAM assessors to carry out an assessment at each of the above stages could be several thousands of pounds per stage.

- **Timing** – many BREEAM credits are affected by basic building form and servicing solutions. Cost-effective BREEAM compliance can only be achieved if careful and early consideration is given to BREEAM-related design and specification details. Clear communication between the client, design team members and in particular, the project cost consultants, is essential.
- **Location** – building location and site conditions have a major impact on the costs associated with achieving very good and excellent compliance.
- **Procurement route** – public–private partnerships and similar procurement strategies that promote long-term interest in building operations for the developer / contractor typically have a position of influence on the building's environmental performance and any costs associated with achieving higher BREEAM ratings.

The Building Regulations 2007 introduced tougher energy and environmental sections, with these new regulations being mandatory from October 2009. In addition the Climate Change Bill will result in Scotland having the most ambitious climate change legislation anywhere in the world with a mandatory target of cutting emissions by 80 per cent by 2050!

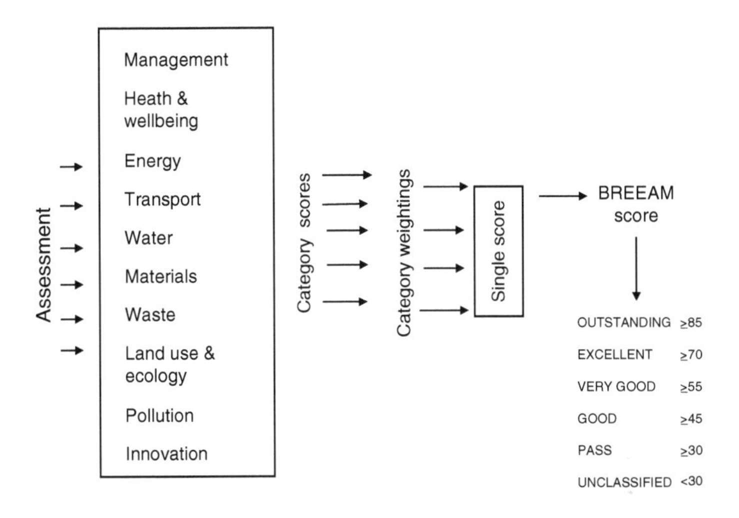

Figure 2.1 BREEAM scoring based on variable and fixed processes

Source: Based on BREEAM SD5078 (2019) Newly constructed buildings.

BREEAM is assessed over several categories, see Figure 2.1. Each category contributors a percentage towards the overall rating

The higher the BREEAM rating the more mandatory requirements there are and progressively harder they become. In 2008 new standards were introduced under BS ISO 15686–5 Service life planning – Buildings and constructed assets.

LEED

LEED is an internationally recognised green building certification system, providing third-party verification that a building or community was designed and built using strategies aimed at improving performance across the following metrics:

- building design and construction,
- interior design and construction,
- operations and maintenance,

- homes, and
- neighbourhood development.

Each of these categories can be awarded one of four LEED designations; platinum, gold, silver and certified.

EcoHomes points / Code for Sustainable Homes

The Code for Sustainable Homes was introduced in 2006 and for a short while was mandatory in England, Wales and Northern Ireland. However, in 2015 the code became advisory only and measures sustainability over the following categories:

- energy and carbon dioxide emissions,
- water,
- materials,
- surface water run-off,
- waste,
- pollution,
- health and wellbeing,
- management, and
- ecology.

VALUE ENGINEERING / MANAGEMENT

Central to the project manager's goal of delivering built assets that meet the functional and operational needs of a client are the techniques of value engineering and value management. Originally known as value engineering (VE), the technique was later rebadged as value management; this approach is now widely practiced in both public and private sectors. SAVE (the International Society of American Value Engineers) defines value engineering as:

> *A powerful problem solving tool that can reduce costs while maintaining or improving performance and quality. It is a function-oriented, systematic team approach to providing value in a product or service.*

It is used at various stages during the project, but the earlier in the process it is introduced, the greater the impact. The basis of value management is to analyse, at the outset, the function of a building, or even part of a building, as defined by the client or end user (see Figure 2.2). Then, by the adoption of a structured and systematic approach, to seek alternatives and remove or substitute items that do not contribute to the efficient delivery of this function, thereby adding value. The golden rule of value engineering / management is that as a result of the value process the function(s) of the object of the study should be maintained, and if possible enhanced, but never diminished or compromised.

The terms in common usage are:

- **Value analysis** – the name adopted by Lawrence D. Miles for his early studies and defined as an organised approach to the identification and elimination of unnecessary cost. As if to emphasise the importance now being placed on value engineering, in 2000 Property Advisors to the Civil Estate (PACE) introduced an amendment to GC/Works/1 – Value Engineering Clause 40(4). The amendment states:

  The Contractor shall carry out value engineering appraisals throughout the design and the construction of the Works to identify the function of the relevant building components and to provide the necessary function reliability at the lowest possible costs. If the Contractor considers that a change in the Employer's Requirements could affect savings, the Contractor shall produce a value engineering report.

- **Value management** – involves considerably more emphasis on problem solving as well exploring in-depth functional analysis and the relationship between function and cost. It also incorporates a broader appreciation of the connection between a client's corporate strategy and the strategic management of the project. In essence, value management is concerned with the 'what' rather than the 'how' and would seem to represent the more holistic approach. The function of value management is to reduce total through-life costs, comprising initial construction, annual operating, maintenance and energy costs and periodic replacement costs, without affecting, and indeed improving, performance and reliability and other required design parameters. It is a function-oriented study and is accomplished by evaluating functions of the project and its subsystems and components to determine alternative means of accomplishing these functions at lower cost.

Using value management improved value may be derived in three predominant manners:

1. Providing for all required functions, but at a lower cost.
2. Providing enhanced functions at the same cost.
3. Providing improved function at a lower cost – the Holy Grail.

Among other techniques, value management uses a value engineering study or workshop that brings together a multidisciplinary team of people. A value engineering study team works under the direction of a facilitator, who follows an established set of procedures, for example the SAVE Value Methodology Standard (see Figure 2.3), to review the project, making sure the team understands the client's requirements and develops a cost-effective solution.

The key player in a VE study is the facilitator or value management practitioner, who must, within a comparatively short time, ensure that a group of people work effectively together. A variety of techniques are used during the study including:

- value trees,
- decision analysis matrices,

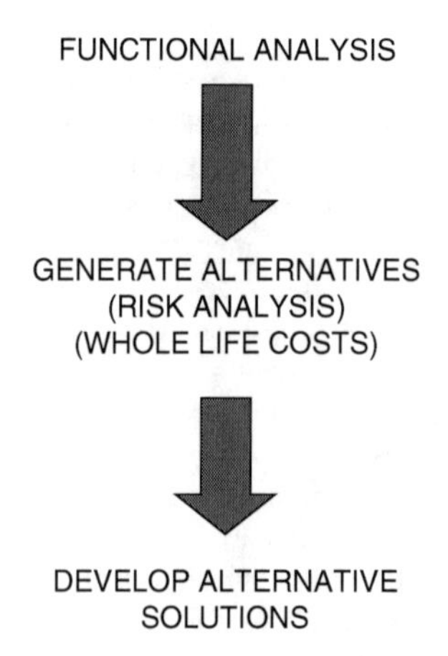

Figure 2.2 Value management

- Functional Analysis System Technique Diagrams (FAST), and
- criteria scoring.

There are numerous variations and adaptations to conducting a value engineering workshop. For example the classic SAVE 40-hour, five-day value engineering workshop (see Figure 2.3). The workshop team is made up of six to eight experts from various design and construction disciplines, who are not affiliated to the project, as it has been found that the process is not so vigorous if in-house personnel are used. In addition, an independent facilitator is recommended as they have proved to be less liable to compromise on the delivery of any recommendations. The assembled team then commences the workshop, following the steps of the SAVE methodology. At the start of the week the group is briefed on the project by key personnel and members of the design and construction team, and the scope of the study is defined. Costs of the project are also carefully examined and analysed using a variety of techniques, as well as compared to other facilities with a similar function. Value management / engineering studies should be linked with both risk and whole life cost assessment (see later in this chapter).

DESIGN

Design management / specialist design / BIM

Design management takes place over the entire project life span. New procurement routes, shorter project timescales, etc. make the management of the design process a critical task for the project manager. The nature and the degree of design management will depend on the procurement path.

Traditional procurement is a linear process, with one process being completed before the next stage can begin. However, other procurement paths, for example design and build, management systems and package deals, allow design, procurement and construction to occur simultaneously. Management contracting, for example, requires that each work package is carried out by a separate work package contractor. The management contractor will be responsible for drawing up the list of packages and there is a case for the lead consultant getting involved / assisting in this process, issuing specifications, clarifying package scope and setting out constraints of packages, for example. Producing a co-ordinated set of designer's drawings and specifications can lead to fewer claims and disputes during the course of the projects.

PRE-STUDY

User/customer attitudes
Complete data files
Evaluation factors
Study scope
Data models
Determine team composition

VALUE STUDY

Information phase
Complete data package
Finalise scope
Function analysis phase
Identify functions
Classify functions
Function models
Establish function worth
Cost functions
Establish value index
Select functions for study
Creative phase
Create quantity of ideas by function
Evaluation phase
Rank and rate alternative ideas
Select ideas for development
Development phase
Benefit analysis
Technical data package
Implementation plan
Final proposals
Presentation phase
Oral presentation
Written report
Obtain commitments for implementation

POST-STUDY

Complete changes
Implement changes
Monitor status

Figure 2.3 Standard 40-hour value engineering methodology

Source: Society of American Value Engineers.

The project manager should ensure that the design of the development is managed in order that:

- the design is completed on time for each stage,
- the design reflects the cost limit and budgets,
- the quality of the design matches the client's aspirations and perceptions, and
- changes are incorporated in a transparent way.

In order to achieve this, the project manager should establish a framework to monitor the design development with the lead designer. Traditionally, the lead designer role and the lead consultant role are undertaken by the same individual, although there may be circumstances where it is appropriate to separate the two roles.

Lead consultant

Traditionally the lead consultant has been the architect (lead designer). What does the lead designer do? Typical responsibilities include:

- co-ordinate preparation of the work stage programme,
- co-ordinate design of all construction elements including work carried out by specialists, suppliers and consultants,
- determine the nature of design outputs and their interface with other factors,
- verify that design,
- liaise with client on any major design matters, and
- establish a design review framework.

Although the lead designer will have overall responsibility for the co-ordination of the design development with other consultants and specialists, it is advisable for the project manager to draw up a design development plan. There may be some resistance from the lead consultant / lead designer to doing this but nevertheless the project manager should persist. The design management plan should consider:

- schedule of information required together with dates,
- BIM data drop dates,
- procedures for introducing design changes,
- monitoring resources,

- responsibilities,
- format of information,
- information exchanges, and
- agreement dates of the value engineering exercise and sustainability checkpoints.

Design co-ordination at each stage of the RIBA Plan of Work (BIM outputs)

Stages 0 & 1 Strategic Definition and Preparation and Brief

DATA DROP 1: MODEL REPRESENTS REQUIREMENTS AND CONSTRAINTS

Advise the client on the purpose and the advantages of using BIM and agree on the level of BIM adoption; e.g. 2D or 3D. Agree intellectual property rights and model ownership and definition of responsibilities. Consider Soft Landings and the scope of BIM including in use issues.
At this stage the model can generate room data sheets illustrating:

- function of the space,
- environmental conditions of the space, and
- finishes.

Stage 2 Concept Design

The focus is on design. The objective at this stage is to decide the scope of the individual studies necessary to develop the project to the completion of the scheme design. This is a stage of intense creative activity and evaluation of alternative strategies for the project. The use and integration of Modern Method of Construction, discussed in Chapter 3, should be considered at this point.

Although during this stage of the design process drawings will be produced, it should not be forgotten that the purpose of this stage is to create a concept, and is concerned with strategy rather than detail. The structural engineer should investigate ground conditions to assess the best possible solution for the substructure which should be co-ordinated with the lead designer. The mechanical and electrical (M&E) consultants should consider the best place for large items of plant, risers, etc. Key steps in this stage include:

- arrange a pre-start meeting,
- commence initial model sharing with design team, and
- identify key model elements.

Stage 3 Design Development

At this stage the main part of the detailed co-ordination work should be undertaken, including:

- BIM,
- data sharing and integration for design co-ordination,
- integration of design components,
- export data for planning,
- technical analysis,
- agree extent of performance specified work, and
- enable design team to access BIM data.

DATA DROP 2: MODEL REPRESENTS OUTLINE SOLUTION

At this stage the model can generate design solutions illustrating:

- function of the space,
- environmental conditions of the space,
- finishes,
- furniture and equipment,
- various schedules,
- tender documentation.

Stage 4 Technical Design

DATA DROP 3: MODEL REPRESENTS CONSTRUCTION INFORMATION

At this stage the model can be used for construction purposes. Various schedules, doors, windows, etc. can be produced. The model is fully co-ordinated technically and fully co-ordinated drawings can be generated. All inputs from the contractor are incorporated into the model:

- data sharing and integration for design co-ordination and detailed analysis including data links between models,

- BIM data used for environmental performance and area analysis,
- export data for planning, and
- data sharing for design co-ordination.

Stage 5 Construction

DATA DROP 4: MODEL REPRESENTS OPERATIONS AND MAINTENANCE INFORMATION

The data being collected at this stage is the operational and detailed functional information supplied by product manufacturers. Particular attention needs to be focused on the needs of the first year of operations including any valid warranties. The model represents the project as built and contains all the information provided by various contractors to maintain it. Information can be extracted from the model that is relevant for FM in order to:

- agree timing and scope of Soft Landings,
- co-ordinate end of construction BIM model data,
- detail modelling, integration and analysis,
- embed specification to model,
- enable access to BIM model for contractors,
- integrate subcontractor performance-specified work model information into BIM model data,
- review construction sequencing with contractor (4D),
- agree timing and scope of Soft Landings, and
- use BIM data for contract administration.

Stages 6 & 7 Handover and Close Out

DATA DROP 5 (AND SUBSEQUENT DROPS): MODEL REPRESENTS POST-OCCUPANCY VALIDATION INFORMATION AND ONGOING O&M

- FM BIM model data issued.
- Study of parametric object information contained within BIM model data.

Design changes

There will be times during the design process when, for any number of reasons, it will be necessary to change the design or form of construction. The project manager should inform the client and other members of the project team at the outset that the earlier that changes can be proposed /

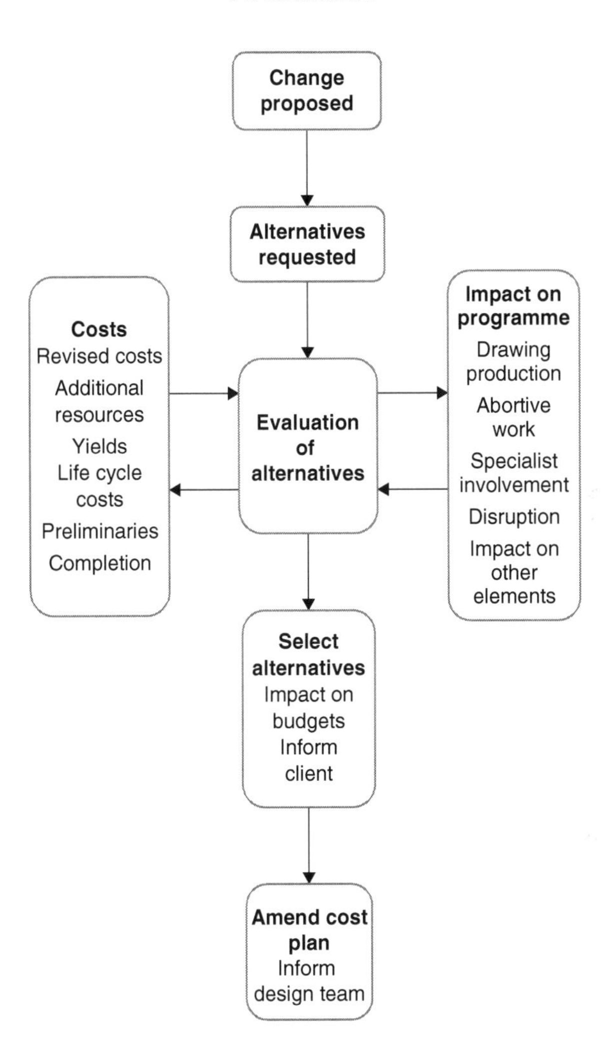

Figure 2.4 Design change process

evaluated and incorporated, the smaller the impact on the project in terms of cost and effectiveness. Figure 2.4 illustrates the design change process. Figure 2.5 illustrates the interaction of timing and consequences of introducing VE into the design process.

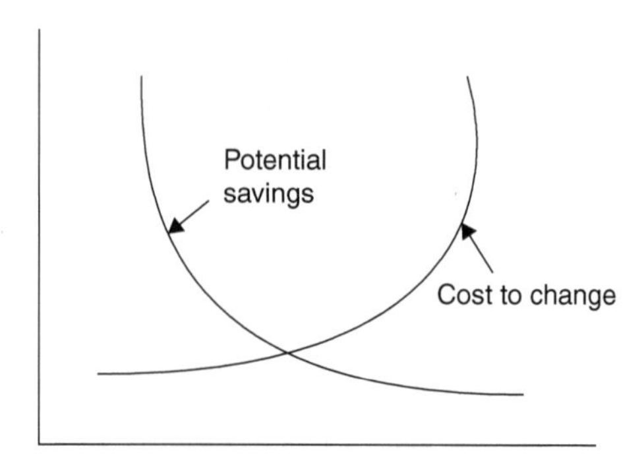

Figure 2.5 Interaction of timing and consequences of introducing VE into the design process

Contractor involvement in the design process

Traditionally, it has been normal practice for designers to appoint specialist subcontractors as part of the design and procurement process typically to prepare the detail design for kitchens shown on the designers general arrangement drawings. However, new procurement routes and forms of contract mean that quite often a contractor can have a significant input into design. It is important for the project manager to appreciate the sometimes fine line between liabilities and obligations that designers have and that a number of design processes need to be considered to manage the associate design risks. In practice much of the contractor design work will be carried out by subcontractors, either domestic or named, and in rare circumstances nominated, although it is the contractor who retains responsibility for design. It may be the case that the type and extent of the works is defined in a performance specification instead of the more traditional prescriptive approach.

Contractor designed portions

The project manager should understand the impact of the design supply chain on procurement and design responsibilities and its need to be successfully integrated into the design process.

The extent of contractor design will vary from project to project but should be determined as early as possible to enable:

- tendering contractors to allow for any design work in their bids, and
- the design to be developed.

Once the extent of the contractor-designed portion has been decided, and the project manager may have their own views on this, and finalised, it should be recorded in a design responsibility matrix. JCT contracts have addressed the need to contractually agree the extent of design work undertaken by specialist contractors (Performance Specified Work) for a number of years. The RIBA Plan of Work 2013 suggests the preparation of a Design Responsibility Matrix (Table 2.4) so that it is clear at the outset to all parties in the project team which aspects of the design will be designed by specialist contractors and which aspects will be constructed on-site from information prepared by the design team. The matrix sets out who is responsible for designing each aspect of the project and when. This document should set out the extent of performance and specified work. The document should be created at strategic level at Stage 1 and fine-tuned to the concept design at the end of Stage 2 in order to ensure that there are no design responsibility ambiguities at Stages 3, 4 and 5.

PLANNING PERMISSION

Currently in the UK planning permission is required for almost all development, including a change of use of an existing building.

The main statutes governing planning law are:

- Town and Country Planning Act 1990 as amended,
- Planning (Listed Building and Conservation Areas) Act 1990,

Table 2.4 Design Responsibility Matrix

Ref.	*Element*
Frame	DC
Doors	CD
Curtain walling	DF
Suspended ceilings	BN
Alarm installation	CV

- Planning and Compensation Act 1991,
- Planning and Compulsory Purchase Act 2004, and
- Localism Act 2011.

The purpose of the planning system is to protect the environment as well as public amenities and facilities. The planning control process is administered by local authorities and exists to *'control the development and use of land and buildings for the best interests of the community'*. The levels of planning are:

- **Regions** – set out regional policy through Regional Planning Guidance.
- **Structure plans** – establish broad planning policies at County Council level.
- **Local plans** – set out detailed policy at District Council level.
- **Neighbourhood plans** – a new right for communities and gives them direct power to develop a shared vision for their neighbourhood and shape the development and growth of their local area.

There are three types of planning permission, all of which are subject to a fee that can range from hundreds to thousands of pounds, depending on the scale of the proposed project:

- **Outline** – this is an application for a development in principle without detail of construction, etc. Generally used for large-scale developments to get permission in principal.
- **Reserved matters** – a follow up to an outline application stage. Reserved matters refers to any of the following in respect of which details have not been given in the application:
  - access,
  - appearance,
  - landscaping,
  - layout, and
  - scale, 'within the upper and lower limit for the height, width and length of each building stated in the application for planning permission…'.
- **Full planning permission** – sometimes referred to as detailed planning permission when a fully detailed application is made. Permission when granted is valid for three years.

In addition there are categories for householder planning consent and listed building consent.

If planning permission is refused then there is an appeals process. From July 2014 a new process was introduced to speed up the process, including:

- faster decision times, with 80 per cent of written representations and hearing appeals to be decided within fourteen weeks, and 80 per cent of non-bespoke inquiries to be decided within twenty-two weeks,
- frontloading the procedures, including submission of a full statement of case by the appellant for all appeals and a draft statement of common ground for hearing and inquiry appeals, and
- shorter timetables for the submission of appeal documents, including earlier notification of interested parties, and the submission of LPA statements and interested party representations within five weeks of the start of an appeal.

Appeals may not be made on the ground of:

- loss of view,
- private issues between neighbours, and
- loss of privacy, etc.

It is thoroughly recommended that prior to a proposed development that the structure plans are read and understood. Buildings erected without planning permission will have a demolition order served on them and the structure will be taken down and destroyed.

Enterprise Zones

Enterprise Zones are government-designated areas in various parts of the UK that offer potential developers valuable tax and business rate breaks as well as simplified planning approval. Once designated, Enterprise Zone status lasts for ten years and development is confined to certain classes of buildings. A new batch of twenty-four Enterprise Zones was announced in 2012 and further extended in 2016/17.

Role of project manager in planning

- If necessary, the project manager should arrange for presentations of the proposed development to local interest groups including any press releases.

- The project manager should also keep informed of the progress of planning applications and advise the client concerning any special conditions imposed by the planners.
- If an element of 'planning gain' is required as a prerequisite for approval, the project manager should explain the financial impact, if any, of this.
- If planning permission is refused then the project manager should seek advice as to whether to amend the proposal or whether an appeal to the Planning Inspectorate has any chance of succeeding. Appeals must be made within the time stipulated.
- If an appeal is to be mounted then the project manager should arrange to brief the specialist consultants required.

The other major statutory consent is building regulation approval.

Building Regulations

Even when planning permission is not required, most building work is subject to the requirements of the Building Regulations. There are exemptions such as buildings belonging to the Crown, the British Airports Authority and the Civil Aviation Authority. Building Regulations ensure that new work and alterations are carried out to an agreed standard that protects the health and safety of people in and around the building. Builders and developers are required by law to obtain building control approval, which is an independent check that the Building Regulations have been complied with. There are two types of building control providers; the local authority and approved private inspectors.

The documents which set out the regulations are:

- Building Act 1984,
- Building Regulations 2010 for England and Wales, as amended,
- Building (Scotland) Act 2003, and
- Building (Scotland) Regulations 2004.

The Building Regulations 2010, England and Wales, are a series of Approved Documents. Each Approved Document contains the Building Regulations' relevant subject areas. This is then followed by practical and technical guidance, with examples, detailing the regulations. The current set of approved documents are in thirteen parts and include details of areas such as; Structural, Fire Safety, Electrical Safety, etc. In Scotland the Approved Documents are replaced with Technical Handbooks.

Contravention of the Building Regulations is punishable with a fine or even a custodial sentence, plus taking down and rebuilding the works that do not comply with the regulations.

There are two approaches to complying with Building Regulations:

1. Full plan application submission – when a set of plans is submitted to the local authority who checks them and advises whether they comply or whether amendments are required. The work will also be inspected as work proceeds.
2. Building notice application – when work is inspected as the work proceeds and the applicant is informed when work does not comply with the Building Regulations.

Once approval is given and a building notice is approved, it is valid for three years.

Party wall issues

There are occasions when party wall disputes have to be addressed. The main types of party walls are:

- a wall that stands on the lands of two (or more) owners and forms part of a building – this wall can be part of one building only or separate buildings belonging to different owners,
- a wall that stands on the lands of two owners but does not form part of a building, such as a garden wall, but not including timber fences, and
- a wall that is on one owner's land but is used by two (or more) owners to separate their buildings.

The Party Wall Act 1996 governs issues arising from the above and covers:

- new building on or at the boundary of two properties,
- work to an existing party wall or party structure, and
- excavation near to and below the foundation level of neighbouring buildings.

This may include:

- building a new wall on or at the boundary of two properties,
- cutting into a party wall,

- making a party wall taller, shorter or deeper,
- removing chimney breasts from a party wall,
- knocking down and rebuilding a party wall, and
- digging below the foundation level of a neighbour's property.

Rights of light

Project managers should be aware that they may be asked to deal with rights of light matters in the following situations:

- an adjoining owner who has concerns regarding a potential infringement to a right of light,
- a developer wishing to assess impacts of rights of light on a development scheme or wishing to determine the maximum size of a potential development,
- the determination of compensation where the parties have agreed that this would be acceptable,
- assessing risk for funders, insurance companies, mortgagees or other interested parties.

Party wall / rights of light issues are specialist areas and expert advice should be sought.

Disability legislation

The main pieces of legislation that govern the adaption of buildings to accommodate people with disabilities are;

- Disability Discrimination Act 2005, and
- Equality Act 2010 in England, Scotland and Wales.

Disability is not always obvious. The Equality Act 2010 defines a person as disabled if '*they have a physical or mental impairment that has a substantial and long term adverse effect on a person's ability to carry out normal day-to-day activities*'.

These requirements must be taken into account in the design of new buildings, as well as existing buildings, and the project manager should be fully aware of the obligations imposed by the legislation. Since the implementation of the Disability Discrimination Act, a company faces prosecution if their premises are inaccessible to people with disabilities and companies must take reasonable steps to ensure that as many disabled people as possible have full access to goods, services and places of interest. Improvements are not restricted to building access, but also include the introduction of additional features such as grab rails, touch-legible signs and visual and audio alarm systems.

Most services are covered by the Disability Discrimination Act and anyone who provides a service to the public or a section of the public is a service provider. There are a few exceptions, for example private clubs that have a meaningful selection process for their members, but in reality most providers of accommodation are service providers, including:

- private landlords,
- housing associations,
- estate agents and managing agents, and
- local authorities providing housing.

A Disability Access Audit will provide a realistic cost-effective action plan.

Health and safety in construction

Construction (Design and Management) (CDM) Regulations 2015

The construction industry traditionally has one of the worst records with regards to the health and safety and wellbeing of its workers. Not unsurprisingly therefore the regulations that relate to health and safety are becoming ever more onerous and the project manager should be aware of their scope. The principal pieces of legislation, developed over a number of years after extensive consultation and partnership between industry and the Health and Safety Executive (HSE), that relate to health and safety are:

- Factories Act 1961,
- Health and Safety at Work Act 1974,
- Personal Protective Equipment at Work Regulations 1992,
- Provision and Use of Work Equipment Regulations 1998,
- 2002 – Revitalising Health and Safety in Construction,
- Work at Height Regulations 2005,
- 2005 – HSC publish Consultation Document with draft Regulations, which combined CDM 94 and Construction (Health, Safety & Welfare) Regulations 1996,

- December 2005 – HSC agreed Regulations should be supported by an Approved Code of Practice and industry-produced guidance,
- Control of Asbestos Regulations 2006,
- Construction (Design and Management) Regulations 2007 (CDM 2007), and
- Construction (Design and Management) Regulations 2015 (CDM 2015).

The introduction of the CDM Regulations in 1994 without doubt led to major changes in how the industry managed health and safety, although several years after their introduction there were concerns from industry and the Health and Safety Executive (HSE) that the regulations were not delivering the improvements in health and safety that were expected. The principal reasons were said to be:

- slow acceptance, particularly amongst clients and designers,
- effective planning, management, communications and co-ordination was less than expected,
- competence of organisations and individuals was slow to improve, and
- a defensive verification approach, adopted by many, which led to complexity and bureaucracy.

The CDM regulations 2015 apply to all construction projects including new build, demolition, refurbishment, extensions, conversions, repairs and maintenance, however, the degree of documentation and supervision required varies according to the size of the project. Notifiable construction work under CDM 2015 are construction projects with:

- construction work lasting longer than thirty days with more than twenty workers working simultaneously at any one point, or
- in excess of 500 person days.

Where more than one contractor is employed the CDM regulations 2015 place the responsibility for the health and safety issues of a commercial construction project on three duty holders:

- **The client** – it is the client's duty to ensure that the project is set up in such a way as to control risks to health and safety to anyone who could be affected. The definition of a client under CDM 2015 is anyone who has construction work carried out for them. The main duty for clients is to make sure their project is suitably managed, ensuring the health and safety of all who might be affected by the work, including members of the public. CDM 2015 recognises two types of client:

- Commercial – defined as an organisation or individual for whom a construction project is carried out in connection with a business, whether the business operates for profit or not. Examples of commercial clients are schools, retailers and landlords.
- Domestic – defined as those having work carried out which is not connected with running a business. Usually, this means arranging for work to be carried out on the property where you or a family member live.

- **The principal designer** manages the pre-construction phase of a project through to the construction phase in addition to liaising with the principal contractor. CDM 2015 requires a principal designer to be appointed when more than one contractor is involved. The principal designer is responsible for:
  - planning, managing and monitoring the pre-construction phase,
  - ensuring that where reasonably practicable, risks are eliminated or controlled through design work,
  - passing information to the principal contractor,
  - ensuring co-operation and co-ordination,
  - ensuring designers comply with their duties,
  - assisting the client in preparing the pre-construction information, and
  - preparing the health and safety file.
- **The principal contractor** is responsible for managing health and safety issues during the construction phase, which involves liaising with the client and principal designer. In the context of CDM 15, for the principal contractor, the term 'manage' includes the following functions:
  - planning the construction phase to minimise risk,
  - managing the construction phase plan,
  - monitoring and revising the plan as necessary,
  - ensuring the site is secure,
  - providing welfare facilities,
  - providing site induction on-site risks, and
  - liaising on the design with the principal designer.

For some projects, as well as smaller domestic-scale projects, two further duty holders are identified:

- the designer, and
- the contractor.

One further group is identified, workers, who are people who work on a construction site under the direction of the contractor and who should be consulted on health and safety issues.

A summary of the roles of the duty holders is shown in Table 2.5, demonstrating that now virtually everyone involved in a construction project has legal duties under CDM 2015.

Two important documents are at the heart of CDM 15:

1. **Construction Phase Plan** – required for all projects.
2. **Health and Safety Plan** – required for project where a principal contractor and designer are appointed, whether domestic or non-domestic.

CONSTRUCTION PHASE PLAN

When drawing up the Construction Phase Plan the following items should be considered for inclusion:

- a general description of the site with key dates and project players,
- the management of the works including site rules, welfare facilities, procedures in the case of fire and medical emergencies,
- timing of meetings,
- how risks specific to the site will be managed, and
- health and safety aims.

Various pro-formas are available on-line from the HSE to help draw up Construction Phase Plans.

HEALTH AND SAFETY PLAN

As mentioned previously a key part of the health and safety process is the Health and Safety Plan and File. A suggested format for a health and safety plan is:

- General introduction.
- Project brief.

Table 2.5 CDM 15 Roles and responsibilities

Commercial clients Organisations or individuals for whom a construction project is carried out that is done as part of a business.	Make suitable arrangements for managing a project, including making sure: • other duty holders are appointed as appropriate, • sufficient time and resources are allocated. Make sure: • relevant information is prepared and provided to other duty holders, • the principal designer and principal contractor carry out their duties, and • welfare facilities are provided.
Domestic clients People who have construction work carried out on their own home (or the home of a family member) that is *not* done as part of a business.	Though in scope of CDM 2015, their client duties are normally transferred to: • the contractor for single contractor projects, or • the principal contractor for projects with more than one contractor. However, the domestic client can instead choose to have a written agreement with the principal designer to carry out the client duties.
Designers Organisations or individuals who, as part of a business, prepare or modify designs for a building, product or system relating to construction work.	When preparing or modifying designs, eliminate, reduce or control foreseeable risks that may arise during: • construction, and • the maintenance and use of a building once it is built. Provide information to other members of the project team to help them fulfil their duties.

Table 2.5 (Cont.)

Principal designers Designers appointed by the client in projects involving more than one contractor. They can be an organisation or an individual with sufficient knowledge, experience and ability to carry out the role.	Plan, manage, monitor and co-ordinate health and safety in the pre-construction phase of a project. This includes: • identifying, eliminating or controlling foreseeable risks, • ensuring designers carry out their duties, Prepare and provide relevant information to other duty holders, Liaise with the principal contractor to help in the planning, management, monitoring and co-ordination of the construction phase.
Principal contractors Contractors appointed by the client to co-ordinate the construction phase of a project where it involves more than one contractor.	Plan, manage, monitor and co-ordinate health and safety in the construction phase of a project. This includes: • liaising with the client and principal designer, • preparing the construction phase plan, • organising co-operation between contractors and co-ordinating their work, Make sure: • suitable site inductions are provided, • reasonable steps are taken to prevent unauthorised access, and • workers are consulted and engaged in securing their health and safety welfare facilities are provided.

Table 2.5 (Cont.)

Contractors Those who carry out the actual construction work, contractors can be an individual or a company.	Plan, manage and monitor construction work under their control so it is carried out without risks to health and safety. For projects involving more than one contractor, co-ordinate their activities with others in the project team – in particular, comply with directions given to them by the principal designer or principal contractor. For single contractor projects, prepare a construction phase plan.
Workers Those working for, or under the control of, contractors on a construction site.	Workers must be consulted about matters which affect their health, safety and welfare.

Adapted from HSE Construction (Design and Management) Regulations 2015.

- Emergency contacts.
- Professional team contacts.
- Project organisation.
- Design risks.
- Site rules and restrictions.
- General arrangements for health and safety.
- Site set-up check list.

Suggested contents for health and safety file:

- as-built drawings and other details,
- risks relating to operation and maintenance,
- information on services and utilities, and
- design criteria and general specification.

Clients should provide those bidding for the work (or those who are preparing to carry out the work) with information about any hazards known about or suspected, for example because they already have information about them in their possession. They should also provide information which can be obtained by sensible enquiries, including surveys and other investigations where necessary. This allows those bidding or preparing for the work to consider these hazards when making their bids or plans, and allows them to allocate resources to control the risks which arise from these hazards.

Dangerous substances

There are many substances involved in the construction process that have the potential to be a danger to health. These include:

- pesticides (agrochemicals, timber treatments, vermin baits) in store or in use,
- lead paints,
- industrial solvents,
- respirable crystaline silica (sand),
- engine exhaust fumes,
- dust and spores from decomposing vegetation, and
- asbestos.

A risk assessment should be undertaken in the form of a Control of Substances Hazardous to Health Regulations (2002) (COSHH) assessment and should be incorporated into the CDM 2015 safety file.

ASBESTOS

According to the HSE asbestos is the single greatest cause of work-related deaths in the UK and was used extensively in building until 1985 when it was banned. No surprise therefore that asbestos-related matters may be one of the most highly regulated issues in the UK. Although the principal regulations were rationalised in 2006, there is still a plethora of approved codes of practice (ACoPs) and official guidance dealing with its use, disturbance, treatment and removal, for example Control of Asbestos Regulations (CAR) (2012).

Examples of how the discovery of asbestos could have a commercial / economic impact are:

- emergency or unplanned stoppage of production and / or cessation of services,
- evacuation of a building, or parts thereof, including the costs of the provision of temporary alternative accommodation and facilities,
- loss of immediate income, due to closure or boycott by customers (e.g. in the case of cinemas, theatres or shops),
- strikes or walkouts by employees or occupants,
- adverse publicity (e.g. for blue-chip companies or schools),
- reduction in value or rental income,
- loss of liquidity of asset (difficulty or inability to sell, lease or license the premises),
- costs of remedial works (removal or treatment and decontamination),
- financial responsibility for injured employees of other parties,
- criminal prosecution (leading to substantial fines and even imprisonment), and
- civil damages for negligence.

During the past thirty years common uses of asbestos in construction have included:

- asbestos cement roofs,
- floor tiles – thermoplastic / PVC tiles,
- insulation board,
- insulation,
- paint,
- roofing felt, and
- Artex decorative coatings.

COST ADVICE / WHOLE LIFE COSTS

There are a number of definitions for whole life costing, but one currently adopted is: '*the systematic consideration of all relevant costs and revenues associated with the acquisition and ownership of an asset*'.

When giving cost advice to the client the project manager should take whole life costs or life cycle costs into account, which includes the consideration of the following cost factors (see also Figure 2.6). The client should be

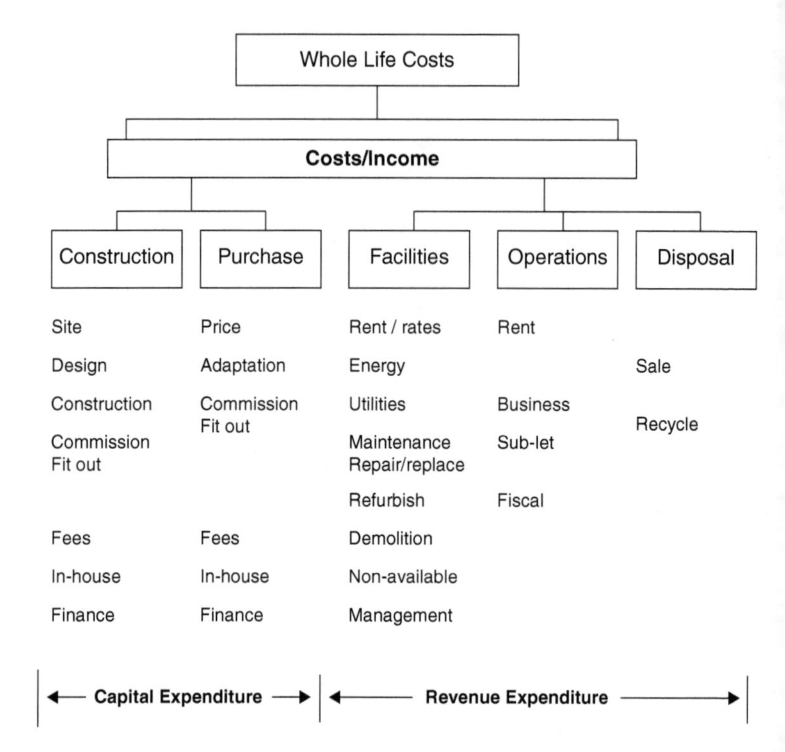

Figure 2.6 Whole life costs

made aware that long term, whole life costs, as well as capital costs, should be considered when decisions are taken.

- Initial or procurement costs, including design, construction or installation, purchase or leasing, fees and charges.
- Future cost of operation, maintenance and repairs, including, management costs such as cleaning, energy costs etc.
- Future replacement costs, including loss of revenue due to non-availability.
- Future alteration and adaptation costs including loss of revenue due to non-availability.
- Future demolition/recycling costs.

Common terms used to describe the consideration of all the costs associated with a built asset throughout its life span are:

- costs-in-use,
- life cycle costs,
- whole life costs, and
- through life costs.

Although whole life costing can be carried out at any stage of the project and not just during the procurement process, its greatest potential for effectiveness is during procurement because:

- almost all options are open to consideration at this time,
- the ability to influence cost decreases continually as the project progresses, from 100 per cent at project sanction to 20 per cent or less by the time construction starts, and
- the decision to own a building normally commits the user to most of the total cost of ownership and consequently there is a very slim chance to change the total cost of ownership once the building is delivered.

Typically, about 75–95 per cent of the cost of running, maintaining and repairing a building is determined during the procurement stage.

Criticisms of whole life costing

Whole life costing is also not an exact science as, in addition to the difficulties inherent in future cost planning, there are larger issues at stake. It is not just a case of asking 'How much will this building cost me for the next fifty years?'. Rather it is more difficult to know whether a particular building will be required in fifty years at all – especially as the current business horizon for many organisations is much closer to three years. Also, whole life costing requires a different way of thinking about cash, assets and cash flow. The traditional capital cost focus has to be altered, and costs thought of in terms of capital and revenue costs coming from the same 'pot'. Many organisations are simply not geared up for this adjustment.

Perhaps the most crucial reason is the difficulty in obtaining the appropriate level of information and data. There is a lack of available data to make the calculations reliable. The RICS Building Cost Information Service operates a subscription-based online service to help estimate future maintenance and operating costs of buildings. The system is dependent on practitioners submitting relevant data for the benefit of others. The increased complexity of construction means that it is far more difficult to predict the whole life cost of built assets. Moreover, if the malfunction of components results in decreased yield or underperformance of the building then this is of concern to the end user / owner. There is no comprehensive risk analysis of building components available for practitioners, only a wide range of predictions of estimated life spans and notes on preventive maintenance – this is too simplistic, there is a need for costs to be tied to risk, including the consequences of component failure. After all, the performance of a material or component can be affected by such diverse factors as:

- quality of initial workmanship when installed on-site and subsequent maintenance,
- maintenance regime / wear and tear – buildings that are allowed to fall into disrepair prior to any routine maintenance being carried out will have a different life cycle profile to buildings that are regularly maintained from the outset, and
- intelligence of the design and the suitability of the material / component for usage – there is no guarantee that the selection of so-called high quality materials will result in low life cycle costs.

Running costs, that is the cost of electricity or gas, are excluded from the RICS Building Running Costs Online service because of their price volatility.

Other commonly voiced criticisms of whole life cost are:

- Expenditure on running costs is 100 per cent allowable revenue expense against liability for tax and as such is very valuable. There is also a lack of taxation incentive, in the form of tax breaks etc., for owners to install energy efficient systems (see later section on capital allowances).
- In the short term, and taking into account the effects of discounting, the impact on future expenditure is much less significant in the development appraisal.

Another difficulty is the need to be able to forecast, a long way ahead in time, many factors such as life cycles, future operating and maintenance costs, and discount and inflation rates. Whole life costs, by definition, deal with the future and the future is unknown. Increasingly, obsolescence is being taken into account during procurement, a factor that it is impossible to control since it is influenced by such things as fashion, technological advances and innovation. An increasing challenge is to procure built assets with the flexibility to cope with changes. Thus, the treatment of uncertainty in information and data is crucial as uncertainty is endemic to whole life costs. Another major difficulty is that the whole life costs technique is expensive in terms of the time required. This difficulty becomes even clearer when it is required to undertake a whole life costs exercise within an integrated real-time environment at the design stage of projects.

In addition to the above changes in the nature of development, other factors have emerged to convince the industry that whole life costs are important.

Whole life cost procurement – critical success factors

- Effective risk assessment – what if this alternative form of construction is used?
- Timing – begin to assess whole life cost as early as possible in the procurement process.
- Disposal strategy – is the asset to be owner-occupied, sold or let?
- Opportunity cost – downtime.
- Maintenance strategy / frequency – does one exist?
- Suitability – matching a client's corporate or individual strategy to procurement.

RICS NRM 3 – order of cost estimating and cost planning for building maintenance works

The final part of the NRM suite, NRM 3 came into effect from 1 January 2015. Following extensive collaboration with the Building Cost Information Service (BCIS), the Chartered Institution of Building Services Engineers (CIBSE) and the Building & Engineering Services Association (B&ES) agreed to adopt the NRM 3 expanded cost structure. This means that the NRM 3 elemental cost structure is now fully aligned with industry standard planned preventative maintenance task schedules and the economic reference life expectancy data structure published by CIBSE Guide M and the BCIS cost analysis. The rationale for concentrating on mechanical services is that in many modern buildings, a large proportion of costs are contained in engineering services, and in addition, reliable data is available for this work.

As previously mentioned, whole life costs have never really been widely adopted in the UK construction industry; however, it is hoped that the RICS new rules of measurement for building maintenance works (NRM 3) will provide a structured basis for measurement of building maintenance works to include the annualised maintenance and life cycle, major repairs and replacements costs of built assets and / or building components that can be expected to accrue during the life cycle of the completed building.

The secondary functions of these rules of measurement for maintenance works include, amongst others, to provide information for:

- input into life cycle cost plans in a structured way so that the same approach is adopted for all life cycle cost plans cash-flows and option appraisals; this in turn will facilitate meaningful comparison and more robust data analysis,
- advising clients on the likely cash flow requirements for the purpose of annual budgeting (and initiating sinking funds) and informing forward maintenance and life cycle renewal plans, and
- informing the implementation of maintenance strategy and procurement stages and cost control of expenditure on maintenance works.

NRM 3 does not deal with operation or occupancy costs, or energy / carbon and environmental costs as these are too unpredictable.

According to the RICS, NRM 3 can be used for:

- initial order of cost estimates during preparation stages,
- cost plans during the design development and pre-construction stages, and
- asset-specific cost plans during the pre-construction phases.

It also offers guidance about:

- the procurement and cost control of maintenance works, and
- the measurement of other items associated with maintenance works that are not included in work items.

For the purpose of developing order of cost estimate, costs in connection with maintenance works, repairs and replacements / renewal works are to be initially ascertained under two separate cost categories as follows:

1. Annual maintenance costs, which are divided into the following sub-categories:
   (a) **Planned preventative costs** – annualised maintenance programme such as preventative maintenance work, includes minor repairs and replacement items (e.g. consumables).

(b) **Reactive costs** – annualised unscheduled programme of responsive maintenance, including minor repairs and replacement items.
(c) **Proactive maintenance** – such as planned inspection of buildings, audits, testing / monitoring regimes and specific operation / management procedures.

2. Forward renewal works costs, which are divided into the following sub-categories:
   (a) **Forecast life cycle renewal plans** – includes cyclical maintenance works (e.g. redecoration and the scheduled major repairs and maintenance works).
   (b) **Unscheduled repair costs** – includes emergency and corrective maintenance.
   (c) **Unscheduled replacement costs** – emergency / corrective maintenance.
   (d) **Improvement and upgrades** – as agreed in the project scope.

The rules for NRM 3 follow the same framework and format as NRM 1: Order of cost estimating and cost planning for capital building works.

The RICS new rules of measurement: Order of cost estimating and cost planning for building maintenance works (NRM 3), is divided into six parts with supporting appendices:

- **Part 1** places order of cost estimating and cost planning in context with the RIBA Plan of Work and the OGC Gateway Process; defines the purpose, use and structure of the rules; clarifies how maintenance relates to other aspects of life cycle costing; defines the cost categories and definitions that constitute the building maintenance works (renewal and maintain); provides preparation rules for defining the brief and project particular requirements; provides guidance on the process of cost estimating and cost planning and levels of measurement undertaken depending on the stage in the building life cycle; advises how to deal with projects comprising multiple buildings or facilities; and explains the symbols, abbreviations and definitions used in the rules.
- **Part 2** sets out the basis for the new rules of measurement for maintenance works by clarifying how maintenance costs relate to construction and life cycle costing; defines the scope and parameters for renewal (R) and maintain (M) cost categories; explains the levels of measurement and process of cost estimating and cost planning – as well the importance of developing a clear and comprehensive employer's maintenance brief and measurement rules.

- **Part 3** describes the purpose and content of an order of cost estimate; defines its key constituents; explains how to prepare and report an order of cost estimate; and sets out the rules of measurement for preparation of order of cost estimates using the floor areas method, functional unit method and elemental method.
- **Part 4** describes the purpose and content of elemental cost planning used for building maintenance works; defines its key constituents; and explains the rules for measurement for the preparation and reporting of formal maintenance cost plans for maintenance works.
- **Part 5** describes the measurement rules for annualised costing of maintenance works; explains the calculation methods used for renewal (R) cost plans generated from capital building cost plans and calculation methods for renewal (R) and maintenance (M) from asset registers and condition surveys and the use of remaining life predicted data.
- **Part 6** comprises the tabulated rules of measurement and quantification of costs of renewal (R) and maintenance (M) works; provides a standardised cost structure for the renewal (R) and maintenance (M) works integrating with the NRM 1 construct (C) cost data structure; methods of codification of maintenance works cost plans; methods of codifying cost plans for works packages; methods of aligning NRM 3 to COBie data structure and definitions for Building Information Models (BIM).
  - Appendix A – Core definition of gross internal floor area (GIFA).
  - Appendix B – Core definitions of net internal area (NIA).
  - Appendix C – Commonly used functional types and functional units of measurement.
  - Appendix D – Special use of definitions for shops.
  - Appendix E – Logic and arrangements for integrating construct (C) to renewal (R) and maintenance (M) works.
  - Appendix F – Maintenance cost categories and definitions and wider life cycle costs.
  - Appendix G – Methods of economic evaluation and discounting equations (time value of money).
  - Appendix H – Information required for determining the maintenance brief and the project particulars.
  - Appendix I – Information required for formal maintenance cost plans.
  - Appendix J – Report template for elemental cost plan for renewal (R) and maintenance (M) – Level 1 codes.

- Appendix K – Report template for elemental cost plan for renewal (R) and maintenance (M) – Level 2 codes.
- Appendix L – Informative example of costing renewal (R) work tasks from capital building works cost plans.
- Appendix M – Informative example of costing renewal (R) work tasks from asset registers / condition surveys.
- Appendix N – Informative example of costing maintenance (M) work tasks from asset registers / task schedules.

RISK

A widely accepted definition of risk is '*an uncertain event or set of circumstances that should it occur, will have an effect on the achievement of project objectives*'.

It is the role of the project manager to deal with the project risks on behalf of the client and to ensure the client's interests are protected when involved in administering, managing, communicating and co-ordinating within the project. Each of the consultants in the development, design team and the organisations in the construction team will be focused on managing their risks on the project. The project manager should take a strategic view on behalf of the client.

One of the most important factors that the project manager has to be able to manage is the potential for risk to impact adversely on project outcomes. Risk has the potential to impact on the development throughout a project's life cycle, from the decision to invest, to procurement, to construction, to running and maintenance costs. Areas that are likely to have a potential for risk are:

- inadequacy of the business case,
- environmental impact,
- disputes and claims,
- economics (macro business cycle),
- late contractor involvement in design process,
- complex contract structures,
- degree of innovation,
- poor contractor capabilities,
- poor management team, and
- poor project intelligence.

The client and the project team's risk viewpoint varies markedly on the importance of the above.

Who carries the risk?

Construction projects carry a great deal of risk and traditionally:

- the project manager is responsible for the identification, analysis and co-ordination of a risk management strategy to ensure development and project risks are minimised and mitigated against,
- the investor / client / owner is responsible for the investment / finance risk,
- the design team consultants are responsible for the design risk,
- the contractor and specialist contractors are responsible for the construction risk, which includes the health and safety of the workforce,
- suppliers and manufacturers are responsible for the performance risk of their components and materials,
- the client / owner is responsible for operating and maintenance risk,
- the insurance industry carries the risk of failure by any of the parties through negligence, accident or force majeure,
- government agencies are responsible for ensuring their codes and regulations set the minimum acceptable standards, and
- maintenance teams and facilities managers take the risk of ensuring the project works in use.

Risk accountability

For each risk it is necessary to consider who is accountable should that risk occur. This person is normally called the risk owner and will be a senior manager or board member. The team must also decide on who can best take responsibility for the action to manage the risk, either on their own or in collaboration with others. This person is normally called the action owner. Individuals, rather than organisations, should be nominated in each case as the latter is too ambiguous. The risk manager should allocate new 'action owners' in the event of people leaving the project team.

Next the team needs to consider what the action owner can undertake to implement one of the strategies outlined above. This will be the management action. Finally, the team needs to decide the date by when the action should be completed and when it should be reviewed. The risk manager should ensure that the team nominate specific dates rather than vague terms such as 'ongoing' or 'next progress meeting'. Poorly defined dates may lead to unmanaged risk escalations and slippage, threatening the successful delivery of the project.

It is the job of the project manager to chase up the action owners in order to ensure that risks are being managed.

In defining the action that the action owner should take, it is necessary to keep things in proportion, assess the resources needed to undertake the action and compare these with the impact should the risk occur. There is little point in expending more resources to manage a risk than would be required were its impact to occur.

The form of contract / procurement strategy will also play a large part in the allocation of risk and many contracts contain provision for risk management to be more transparent.

The questions that should be addressed by the project manager are:

- What are the risks?
- What will their impact be?
- What is the likelihood of the risks occurring?

The processes involved with risk management are:

- risk analysis, and
- risk management.

How to deal with risk

- **Avoidance** – take proactive actions to stop the risk from happening.
- **Contingency** – this approach involves taking the decision to let the risk happen and making contingency plans to absorb the action. It can take the form of:
    - A strategic contingency, i.e. having a plan B – in the event of the risk occurring then implement a pre-planned alternative.
    - Cost contingency – have a reserve of uncommitted cash to meet the cost of covering the financial consequences should the risk occur. Usually allowed for as a percentage of the total cost. It should be transparent.
    - Time contingency – allow for some contingency time in the programme should subcontractors or materials not turn up on time.
    - Taking out insurance.
- **Mitigation** – action to reduce the probability of the risk occurring or, if the risk does occur, to mitigate the impact should the risk occur:
    - minimise the probability that the risk will occur, and

  - minimise the impact should the risk occur ensure that the impact is kept to a minimum.
- **Transfer** – agree at the start of the contract who is going manage the risk and transfer risk to those who are best able to manage it.
- **Take no action** – take a positive decision to ignore risk on the basis that the chances of risk impacting on the project is minimal, as is the potential cost set against the high cost of trying to manage risk.

It is never too early to start considering risk! The management of risk should be a continuous process not just to be considered at the start of a project and then forgotten; it must be constantly revisited during the duration of the project.

The success of project risk management is dependent on the effective implementation of the risk responses. The objectives of the risk monitoring and control processes are to:

- review on a monthly basis the current risk profile and identify changes in the risk probabilities and impacts,
- monitor on a monthly basis the implementation of risk responses and implement any necessary changes,
- update on a quarterly basis the risk register with any new risks and associated responses based on changes in project scope, project progress and changing risk generators, and
- review on a quarterly basis the level of project risk management maturity of each project in the programme.

Risk attitude

The project manager needs to be aware that every organisation / client will have a different perception of risk. Risk-loving, risk-neutral and risk-averse organisations will respond differently to the same risk but, there is no scientific way to measure perception or attitude that can be used in risk analysis. Hence, some risks will be over-compensated, while others will be underestimated.

Risk management

The aim of risk management is to ensure that risks are identified at project inception, their potential impacts allowed for and, where possible, the risks or their impacts minimised.

Risk identification

Successful risk management depends on accurate risk identification. Both management practice and engineering techniques should be applied to determine how things might go wrong. When identifying potential risks, it is important to distinguish between the origin of a risk and its impact.

Risk assessment

The purpose of risk assessment is to understand and quantify the likelihood of occurrence and the potential impacts on the project outcome. Various analytical techniques are available, but the key features are:

- qualitative assessment – to describe and understand each risk and gain an early indication of the more significant risks, and
- quantitative assessment – to quantify the probability of each risk occurring and its potential impact in terms of cost, time and performance.

Qualitative assessment

A descriptive written statement of relevant information about a potential risk should be prepared. Issues to be considered should include:

- the stages of the project when it could occur,
- the elements of the project that could be affected,
- the factors that could cause it to occur,
- any relationship or inter-dependency on other risks,
- the likelihood of it occurring, and
- how it could affect the project.

Quantitative assessment

The likelihood of a risk occurring is given a numerical probability. This is measured on the following scale:

0 = impossible for risk to occur,
0.5 = even chance of risk occurring, and
1 = risk will occur.

Possible consequences of a risk arising is quantified in terms of:

- cost – additional cost, above the base estimate for the project outturn,
- time – additional time, beyond the base estimate of the completion date for the project, and
- performance – the extent to which the project would fail to meet the user requirements for standards and performance.

Risk monitoring and control

The aim of risk management is to minimise the opportunity for risks to occur and their impacts should they occur. There are various options available when evaluating the risk response strategy. Care should be taken when considering the management actions available to ensure that the potential impact of each risk is not outweighed by the direct costs to the department from:

- the cost of reducing the risk,
- the cost of transferring the risk (or the cost of insurance), and
- all management and administrative time, consultants' fees and other charges associated with managing and dealing with the risk.

For each project, a risk management plan should be prepared and updated regularly to summarise the risk management process to date.

Risk response

A risk response should only be decided after its possible causes and effects have been considered and fully understood. It will take the form of one or more of the following management actions:

- avoidance,
- reduction (including elimination),
- transfer, or
- retention (including sharing).

As a general rule, risks should be allocated to those best placed to manage them.

Risk avoidance

Where risks have such serious consequences on the project outcome to make them totally unacceptable in the context of the client's internal rules or the project objectives, risk avoidance measures might include a review of the project objectives and a reappraisal of the project, perhaps leading to the replacement of the project or its cancellation.

Risk reduction

Typical action to reduce risk can take the form of:

- redesign – including that arising out of VE studies,
- more detailed design or further site investigation – to improve the information on which estimates and programmes are based,
- different materials or permanent equipment – to avoid new technology, unproven systems or long delivery items,
- different methods of construction – to avoid inherently risky construction techniques,
- changing the project execution plan – to package the work content differently, or
- changing the contract strategy – to allocate risk between the project participants in a different way.

Risk reduction measures lead to a more certain project outcome. They usually result in a direct increase in the base estimate, and a correspondingly greater reduction in risk allowance.

Risk transfer

Where accepting a risk would not result in best value for money, it could be transferred to another party, who would be responsible for the consequences should the risk occur. The object of transferring risk is to pass the responsibility to another party better able to control it. Risk transfer is usually from:

- the client to a design consultant,
- the client to a contractor,
- the contractor to a subcontractor,

- the client or other parties to an insurer in the form of insurance cover, or
- the contractor or the subcontractor to a bank or a surety in the form of warranties, bonds and guarantees.

Whenever a risk is transferred to another party a premium is usually paid. This results in a direct increase in the base estimate and a reduction in risk allowance. To provide value for money, risk transfer should only be carried out where the overall potential cost of the risk to the department is reduced by more than the cost of the premium.

Factors that should be considered include:

- Who is best able to control the events which may lead to the risk occurring?
- Who can control the risk if it occurs?
- Is it preferable for the client to be involved in the control of the risk?
- Who should be responsible for a risk if it cannot be controlled?
- If the risk is transferred to a project participant:
    - Is the total cost to the client likely to be reduced?
    - Will the recipient be able to bear the full consequences if the risk occurs?
    - Could it lead to different risks being transferred back to the client?
    - Would the transfer be legally secure – will the transfer be accepted under common law?

Risk retention

Risks that are not transferred or avoided are retained by the client although they may have been reduced or shared. These risks must continue to be managed by the client to minimise their potential impact.

Risk and procurement strategies

Risk and procurement strategies are interrelated. The chosen strategy and the forms of contract influence the allocation of risk, the project management requirements, the design strategy, the employment of consultants and contractors, and the way in which the client's project team and the various designers, consultants, contractors and suppliers work together.

Tools and techniques of risk identification

There are many tools and techniques available to the project manager. Some of the more traditional and widely used are:

- Ishikawa (fishbone) diagram,
- risk register, and
- decision tree.

Ishikawa diagram

The Ishikawa diagram (Figure 2.7), also known as the fishbone diagram or the cause-and-effect diagram, is a tool used for systematically identifying and presenting all the possible causes of a particular risk in graphical format. The possible causes are presented at various levels of detail in connected branches, with the level of detail increasing as the branch goes outward, i.e., an outer branch is a cause of the inner branch it is attached to. Thus, the outermost branches usually indicate the root causes of the problem.

The Ishikawa diagram resembles a fish skeleton, hence the name fishbone diagram. It has a box, the fish head, that contains the statement of the problem at one end of the diagram. From this box originates the main branch the fish spine of the diagram. Sticking out of this main branch are major branches that categorise the causes according to their nature. Experienced

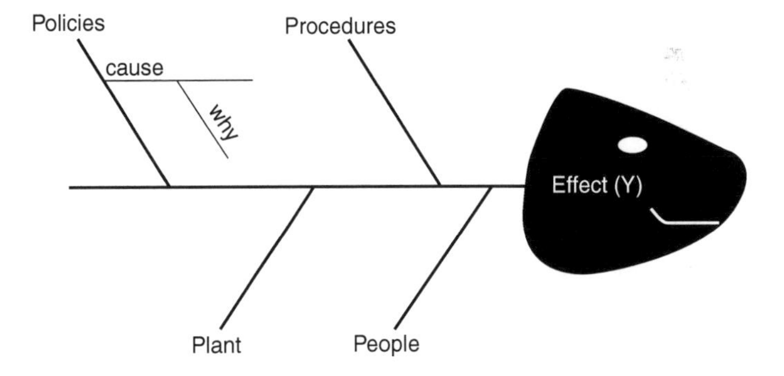

Figure 2.7 The Ishikawa diagram (fishbone diagram)

users of the diagram add more branches and / or use different categories, depending on what would be more effective in dealing with the risk.

The fishbone diagram approach is one way to capture the different ideas of the project team and stimulate the team's brainstorming on root causes on the cause and effect diagram. The fishbone will help to visually display the many potential causes for a specific problem or effect. It is particularly useful in a group setting and for situations in which little quantitative data is available for analysis.

To construct a fishbone, start with stating the problem in the form of a question, such as 'Why is ...?' Framing it as a 'why' question will help in brainstorming, as each root cause idea should answer the question. The team should agree on the statement of the problem and then place this question in a box at the head of the fishbone.

The rest of the fishbone then consists of one line drawn across the page, attached to the problem statement, and several lines, or bones, coming out vertically from the main line. These branches are labeled with relevant categories. Once the branches labeled 'brainstorming' can produce possible causes, attach them to the appropriate branches. For each cause identified, continue to ask 'Why does that happen?' and attach that information as another bone of the category branch to arrive at the true drivers of a problem.

Risk register

A risk register can either be a standard pro-forma or bespoke item drawn up especially for a project as a result of brainstorming.

A risk register will generally:

- identify risks that are capable of being identified,
- assess the probability of the risk occurring,
- develop a range of possible outcomes (worst case, medium case and best case) for each risk,
- value the outcome and the timing of each risk,
- assign probabilities to each outcome,
- calculate the expected value of each risk as the weighted average value of probability of the risk occurring, the outcome of the values and their probabilities, and
- finally value each risk.

See Table 2.6 for a generic risk list. A sample risk list is included in Appendix D.

Table 2.6 Generic project sample risk list

Technical, quality, or performance risks

Examples include reliance on unproven or complex technology, unrealistic performance goals, long-term performance, process roadblocks, new emerging initiatives, increases in complexity, etc.

External risks

Examples include a shifting regulatory environment, labour issues, changing customer priorities, government agency risks and weather. Also to be considered are consultant and vendor contract risks, contract type and contractor responsibilities.

Organisational risks

Examples include lack of prioritisation of projects, inadequacy or interruption of funding, inexperienced and poorly developed and trained workforce, and resource conflicts with other projects in the organisation.

Project management risks

Examples include poor allocation of time and resources, inadequate quality of the project plan, lack of project manager delegated authority, and lack of project management disciplines.

Decision trees

The project manager should be aware that there are often interrelationships between risks, sometimes referred to as consequential risks, which increase the complexity of trying to assess them. It is not uncommon for one risk to trigger or increase the impact and / or likelihood of another. Such knock-on events can turn a relatively minor event, for example the redecoration of a room, in to a major event by holding up the completion and handover of the building.

Decision trees allow the project manager and project team to map options and their costs against the probability of them occurring. In Figure 2.8 two possible suppliers are compared. The cost will depend on when the suppliers are able to deliver as indicated on the example. Based on their track record, an estimation of how they will perform is entered into the equation (based on .1–1.0). By multiplying the basic price plus pain / gain payments the various overall costs can be predicted. The example in Figure 2.8

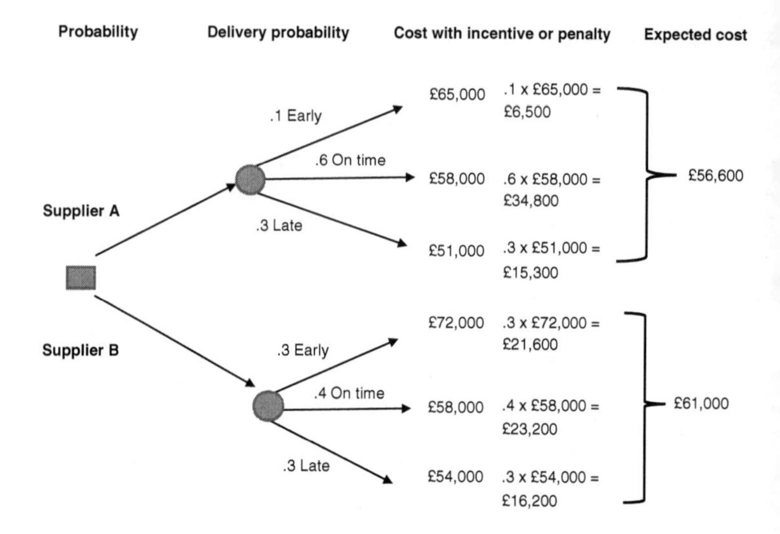

Figure 2.8 Decision tree example

suggests that Supplier A would charge £56,600 and has a 70 per cent chance of delivering early or on time, whereas Supplier B would charge £61,000 and would also have a 70 per cent chance of delivering early or on time. Therefore, Supplier A will be the better option.

Other risk identification techniques include:

- brainstorming,
- cause and effect diagrams,
- SWOT analysis,
- post-project reviews / lessons learned,
- questionnaires,
- project documentation review,
- sensitivity analysis.

Software packages

A number of software packages are available including @RISK, which performs risk analysis using Monte Carlo simulation to show the project manager many possible outcomes in an Excel spreadsheet – and illustrates how likely they are to occur. This means it is possible to judge which risks to take and which ones to avoid, allowing for the best decision-making under uncertainty. With @RISK, you can answer questions like, 'What is the probability of profit exceeding £1 million?' or 'What are the chances of losing money on this venture?'

Qualitative risk analysis

The purpose of qualitative analysis is to prioritise the risks in terms of importance, without quantifying (costing) them. An assessment is made of the likelihood that the risk will occur and the magnitude of its potential impact. The qualitative severity rating is arrived at by multiplying the likelihood of occurrence by the qualitative impact. Likelihoods and impacts can be categorised as follows:

Likelihood		*Probability*
5.	very high – almost certain to occur	75–99%
4.	high – more likely to occur than not	50–74%
3.	medium – fairly likely to happen	25–49%
2.	low – but not impossible	5–24%
1.	very low – unlikely	4–0%

Impact on project costs		*Probability*
5	very high – critical impact on cost	2%
4.	high – major impact on cost	1.50%
3.	medium – reduces feasibility	1.00%
2.	low – minor loss	0.50%
1.	very low – minimal loss	0.25%

Therefore, if a medium likelihood (3) is multiplied by a high impact on cost (4), this gives a total rating of 12.

Sensitivity analysis

A sensitivity analysis is a simple tool that the project manager can use in order to demonstrate to a client the potential impact of risk on the project outcomes and answer the 'what if' question. The method consists of changing input variables by predicted magnitudes and recording the changes in model outputs.

- When only one variable changes at a time, the output changes linearly so only two points are generally needed on either side of the expected value (usually in percentage terms for comparison with other variables – best and worst scenarios).
- If multiple input variables are analysed, one can determine which input variables affect the outcome to a larger degree (the variable with the larger slope, unless plotted horizontally).
- Knowing the sensitivity of the model to various inputs can better inform decisions and help determine if more accurate input information is needed.
- Graphical representations such as spider diagrams and tornado charts help demonstrate the sensitivity of input variables.

Cost–benefit analysis (CBA)

For public sector projects the project manager may have to consider using cost–benefit analysis (CBA). CBA is used mainly for high profile large public sector projects that have mainly quantifiable economic benefits against potential impacts on the environment.

CBA attempts to set the value of the benefits against the costs of the development and has been used on projects such as the 2012 Olympic Games and Heathrow Terminal 5. The trouble with CBA is that the range of benefits and costs can be so wide, and in some cases so subjective, that it can be difficult to try and evaluate their worth. CBA presumes that a monetary value can be assigned to each project input (costs) and each output (benefit) resulting from the proposed project. The value of the costs and benefits are then compared and, in basic terms, if the benefits exceed the costs the project is deemed to be worthwhile. It may be that hybrid forms of CBA may be used with other models such as business cases, etc. in order to try and obtain development approvals.

PROCUREMENT STRATEGIES

Procurement may be thought of as *obtaining goods and services*. There are a number of alternative procurement strategies available which reflect the importance to the client of compliance with certain parameters (Figure 2.9).

The choice of procurement will affect contractual relationships and also the distribution of risk between the client and contractors / subcontractors (Figure 2.10).

- **Procurement strategy** identifies the best way of achieving the objectives of the project and value for money, taking account of the risks and constraints, leading to decisions about the funding mechanisms and asset ownership for the project. The aim of a procurement strategy is to achieve the optimum balance of risk, control and funding for a particular project.
- **Procurement route** delivers the procurement strategy. It includes the contract strategy that will best meet the client's needs. An integrated procurement route ensures that design, construction, operation and maintenance are considered as a whole; it also ensures that the delivery team work together as an integrated project team.

The project manager should discuss with the client his / her attitude to the above parameters including their willingness to accept risk.

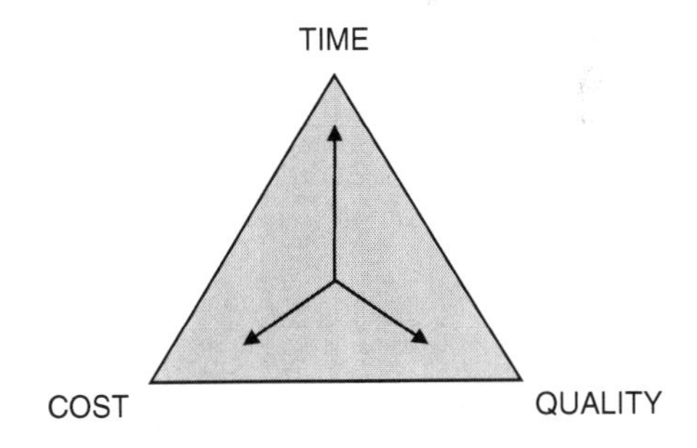

Figure 2.9 Procurement drivers

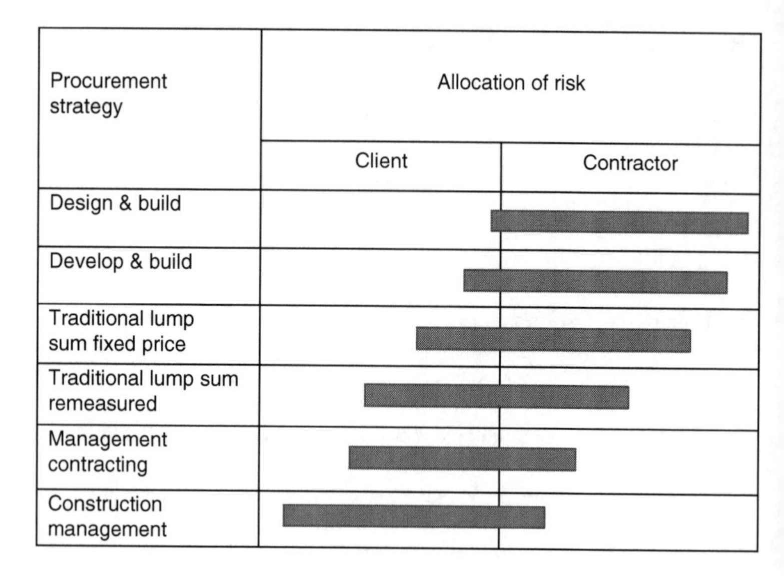

Figure 2.10 Allocation of procurement risk

Generally, there are two broad strategies for obtaining a bid, these are by:

- negotiation, or
- competition.

Negotiation

Negotiation involves the client and contractor's representatives sitting down and negotiating a price for a project without the benefit of competition from other contractors. It is viewed with suspicion by many who consider that, without competition, a contractor will take advantage of the situation and negotiate a higher than market price as the client has no alternative other than to accept it. However, the advantage of negotiation is that the estimating / bidding process can be shorter than with the competitive approach and that if there is trust between the parties the tender can be no more costly than by introducing competition. Due to the potential to deliver a project earlier than otherwise would have been the case, project finance may be recouped earlier and finance charges reduced. In this situation the estimator will be involved in providing the negotiator with data on material, labour and plant costs, etc.

Competition

The majority of work in the construction industry is won through competition, with three or four contractors or subcontractors submitting confidential bids; it's a system that always nearly guarantees that the lowest price wins. The most popular procurement routes that use competition are:

- single-stage competitive tendering,
- two-stage competitive tendering,
- design and build and variants, and
- management.

PROCUREMENT ROUTES

Single-stage selective tendering

The chief characteristics of traditional single-stage competitive tendering are:

- It is based on a linear process with little or no parallel working, resulting in a sometimes lengthy and costly procedure.
- Competition or tendering cannot be commenced until the design is completed.
- The tender is based on fully detailed bills of quantities.
- Design and technical development are carried out by the clients' consultants and do not involve the contractor, unlike some other strategies described later.

Other procurement paths have attempted to shorten the procurement process with the introduction of parallel working between the stages of client brief, design, competition and construction.

The advantages of single-stage competitive tendering are:

- it is well known and trusted by the industry,
- it ensures competitive fairness,
- for the public sector, it allows audit and accountability to be carried out, and
- it is a valuable post-contract tool that makes the valuation of variations and the preparation of interim payments easier.

The disadvantages are:

- a slow sequential process,
- no contractor or special involvement, and
- pricing can be manipulated by tenderers, and can be expensive.

Two-stage competitive tendering

First used widely in the 1970s, this process is based on the traditional single-stage competitive tendering – bills of quantities and drawings are used to obtain a lump sum bid. Advantages include early contractor involvement, a fusion of the design / procurement / construction phases and a degree of parallel working that reduces the total procurement and delivery time. A further advantage is that documentation is based upon bills of quantities and therefore should be familiar to all concerned. Early price certainty is ruled out, as the client can be vulnerable to any changes in level in the contractor's pricing between the first and second stages.

Unless the parameters of the project have altered greatly there should be no significant difference between the stage one and stage two prices. Once a price is agreed a contract can be signed and the project reverts to the normal single-stage lump sum contract based on firm bills of quantities, however the adoption of parallel working during the procurement phase ensures that work can start on-site much earlier than the traditional approach. Also the early inclusion of the main contractor in the design team ensures baked-in buildability and rapid progress on-site.

Design and build (D&B) and variants

Design and build (D&B), or design and construct, is a generic term for a number of procurement strategies where the contractor both designs and carries out the works. This approach is extensively used in France where both contractors and private practices are geared up to provide this service to clients. In the UK this approach has only become common during the last thirty years or so. The various forms of design and build are as outlined below.

Traditional design and build

The contractor is responsible for the complete design and construction of the project. D&B is one of the procurement systems currently favoured by many public sector agencies and private sector clients because:

- D&B gives a client the opportunity to integrate, from the outset, the design and the construction of the project,
- the client enters into a single contract with one company, usually a contractor, who has the opportunity to design and plan the project in such a way as to ensure that buildability is baked in to the design,
- with specialist involvement from the start, this approach promises a shorter overall delivery time and better cost certainty than traditional approaches,
- the total delivery speed of D&B compared with traditional approaches can be 30–33 per cent faster,
- the percentage of projects that exceeded the original estimate by more than 5 per cent is approximately 20 per cent in B&D compared to 32 per cent for traditional procurement, and
- D&B is recommended by the Cabinet Office for procurement within a partnership arrangement.

The main criticisms of D&B procurement are centred around the lack of control over quality of design, with little time being allocated for design development and possible compromises over quality to provide cost savings by the contractor. It is possible for the client to employ independent professional advice to oversee a D&B contract.

Successful use of D&B relies on the contractor preparing proposals that include:

- a contract sum analysis that itemises the financial detail on an elemental basis, and
- detailed proposals of how the requirements of the client's brief will satisfied.

When D&B is chosen as the procurement route the contractor will be responsible for both design, estimating and building the project. It is unlikely that a bill of quantities will be prepared, instead the contractor will prepare a number of work packages to be priced by subcontractors in the order required by the project. The process gives more latitude to the contractor to manage the process in a way that maximises profit and delivers the project in the shortest possible time.

Other variants of D&B are:

- **Enhanced design and build** – the contractor is responsible for the design development and working details, as well as construction of the project.

- **Novated design and build** – the contractor is responsible for the design development, working details and supervising the subcontractors, with assignment / novation of the design consultants from the client. This means that the contractor uses the client's design as the basis for their bid.
- **Package deal and turnkey** – the contractor provides standard buildings or system buildings that are in some cases adapted to suit the client's space and functional requirements.

Management procurement

During the 1970s, and particularly the 1980s, commercial clients and property developers started to demand that projects were procured more quickly than had been the case with single-stage selective tendering. The three main management systems are:

- management contracting,
- construction management, and
- design and manage.

With fast track methods the bidding and construction phases are able to commence before the design is completed and there is a degree of parallel working as the project progresses. This obviously is high risk as the whole picture is often unknown at the time the works commence on-site. This risk is exacerbated when this strategy is used for particularly complex projects or refurbishment contracts.

Management contracting

Management contracting is not only popular with developers, as projects are delivered more quickly, but also contractors, as the amount of exposure to risk for them is substantially lower than other forms of procurement. This is because a management contractor only commits the management expertise to the project, leaving the actual construction works to others. Management contracting was first widely used in the 1970s and was one of the first so-called fast track methods of procurement that attempted to shorten the time taken for the procurement process. When this procurement method is adopted, the client's quantity surveyor will prepare a number of work package bills of quantities to be priced by subcontractors.

Procurement begins with the selection of a management contractor; as the management contractors' role is purely to manage, it is not appropriate to appoint a contractor using a bill of quantities. Selection therefore is based on the service level to be provided, the submission of a method statement and the management fee (expressed as a percentage of the contract sum). This can be done on a competitive basis. As the management contractor's fee is based on the final contract sum, there is little incentive to exert prudence.

The advantages are:

- work can start on-site before the design work is complete,
- earlier delivery of project and return on client's investment, and
- the client can have a direct link with package contractors (see below).

In order to provide a degree of protection for the client a series of collateral warranties can be put in place.

The disadvantages are:

- high risk for client,
- firm price in not known until final package is let,
- difficult for the quantity surveyor to control costs, and
- any delay in the production of information by the design team can have disastrous consequences on the overall project completion.

A distinct JCT form of contract exists for management contracts.

Construction or contract management

Construction or contract management is similar in its approach to management contracting in as far as the project is divided into packages. However, the construction manager adopts a consultant's role with direct responsibility to the client for the overall management of the construction project, including liaising with other consultants. Construction managers are appointed at an early stage in the process and, as with management contracting, reimbursement is by way of a pre-agreed fee. Each work package contractor has a direct contract with the client, this being the main distinction between the above two strategies.

Design and manage

When this strategy is adopted a single organisation is appointed to both design the project as well as manage the project using work packages. It is an attempt to combine the best of design and build and management systems. The characteristics are:

- a single organisation both designs and manages,
- the design and management organisation can be either a contractor or a consultant,
- work is let in packages with contracts between the contractor or client, dependent on the model adopted, and
- reimbursement is by way of an agreed fee.

Cost reimbursement contracts

This group of procurement strategies reimburse the contractor for the actual cost of carrying out the works; labour, materials and plant, plus an agreed percentage to cover overheads, profit and other costs.

Cost-plus contracts

Cost-plus contracts are best used for uncomplicated, repetitive projects such as road contracts. The system works as follows:

- The contractor is reimbursed on the basis of the prime cost of carrying out the works, plus an agreed percentage to cover overheads and profit. This can be done by the contractor submitting detailed accounts for labour, materials and plant that are checked by the quantity surveyor.
- Once agreed the contractors costs are added. There is no tender sum or estimate. The greater the cost of the project, the greater the contractor's profit.
- The estimator has little to do in this method of procurement apart from calculate the percentage addition.

Contractor designed portion

Increasingly it has been the case that portions of the responsibility for the design of a project are handed over to the contractor – the contractor designed portion.

Contractor designed portion (sometimes referred to as 'contractor's design portion' or CDP) can be incorporated within JCT construction contracts. It is an agreement for the contractor to design specific parts of the works. The contractor may in turn subcontract this design work to specialist subcontractors. This should not be confused with design and build contracts where the contractor is appointed to design the whole of the works. Until 2005, the JCT Standard Building Contract had a contractor's designed portion supplement for use where the appointed contractor was required to design specific parts of the works, but provision was incorporated into the JCT (16). The JCT Intermediate Building Contract and the JCT Minor Works Contract have a 'with contractor's design' option and a separate subcontract with subcontractor's design.

The client's requirements for contractor's design will generally be set out in the tender documents as Employer's Requirements in response to which the contractor will submit Contractor's Proposals. The main contractor may be given the responsibility for ensuring that its subcontractor's design is submitted for and obtains Building Regulations approval. The client's design team will usually submit specialist design to the planning authorities when necessary to satisfy planning conditions since it is generally the design team's responsibility to obtain full planning consent. The introduction of Building Information Modelling (BIM) is likely to have an impact on contractor's designs. If at the time of the main contractor's appointment, a fully integrated and collaborative building model has been prepared, it is unlikely that this will simply be handed over to the main contractor. The model is more likely to remain under the management of the client's design team. Whoever is managing the model will be the party that has to manage the collective input of specialist designers.

Pre-qualification questionnaires

Increasingly contractors and subcontractors are required to pre-qualify in order to be placed on a framework or list of selected firms. There follows a suggested format for a pre-qualification process.

Stage one of the pre-qualification process has the following objectives:

- limiting tendering to contractors with the necessary skills and experience to successfully complete the project;
- avoiding unnecessary cost to industry in the preparation of expensive tenders that have limited chance of success, and
- ensuring a competitive tender process, leading to a best value for money outcome for the client.

To achieve this, a six-stage process can be used:

1. Preparation of pre-qualification documents, including a project brief to detail the requirements of the project.
2. Advertise and issue pre-qualification documents to interested parties. The documents will require parties to demonstrate various financial, managerial and technical skills in addition to an appreciation for the project. The documents will contain the evaluation criteria, the evaluation procedures and the proposed timing of the evaluation process. Evaluation criteria will be chosen to allow the evaluation team to determine the most suitable parties to be invited to tender.
   Criteria may typically include:
   (a) financial status,
   (b) legal status (entity),
   (c) relevant experience,
   (d) available resources (staff, plant, subcontractor and supplier relationships),
   (e) performance history, including safety and quality claims
   (f) demonstrated understanding of the project and associated significant issues including technical, environmental and community.
3. A briefing will be held, at which interested parties will be briefed on the particulars of the project and where parties may ask questions.
4. After receipt of pre-qualifications, submissions should be comparatively assessed in accordance with the evaluation criteria.
5. The evaluation team may seek clarification of any issues from applicants, verbally or in writing, but may not solicit additional information.
6. A list of pre-qualified tenderers is published. Successful and unsuccessful parties should be invited to an individual debrief. When establishing the number of tenderers to be included within the select list, clients should consider the competing aims of:
   (a) the cost to industry of the preparation of the tender and the possibility of success for any particular tenderer, and
   (b) ensuring a competitive tender field.

Unnecessary and / or irrelevant information and unnecessary copies should not be sought from parties seeking to pre-qualify and should not be supplied.

Target cost

A variant of cost-plus contracts, this strategy incentivises the contractor by offering a bonus for completing the contract below the agreed target cost. Conversely, damages may be applied if the target is exceeded.

Term contracts / schedule of rates

This approach is suitable for low value repetitive works that occur on an irregular basis. Contractors are invited to submit prices for carrying out a range of items based on a schedule of rates. Contractors are required to quote a percentage addition on the schedule rates. Used extensively for maintenance and repair works.

Negotiated contracts

This strategy involves negotiating a price with a chosen contractor or contractors, without the competition of the other methods. Generally regarded by some as a strategy of last resort and an approach that will almost always result in a higher price than competitive tendering, it has the following advantages:

- an earlier start on-site than other strategies, and
- the opportunity to get the contractor involved at an early stage.

The contractors selected for this approach should be reputable organisations with a proven track record and the appropriate management expertise.

Partnering / frameworks

Many of the established procurement paths have a reputation of perpetuating the 'them and us' culture, and in an attempt to establish a more collaborative approach to construction, alternative procurement paths have been established.

Partnering

Partnering relies on co-operation and teamwork, openness and honesty, trust, equity and equality between the various members of the supply chain.

Partnering is:

- a process whereby the parties to a traditional risk transfer form of contract (i.e. the client, the contractor and the supply chain) commit to work together with enhanced communications, in a spirit of mutual trust and respect towards the achievement of shared objectives,
- a structured management approach to facilitate teamwork across contractual boundaries that helps people to work together effectively in order to satisfy their organisations' (and perhaps their own) objectives,
- a means of avoiding risks and conflict – there isn't one model partnering arrangement; it is an approach that is essentially flexible, and needs to be tailored to suit specific circumstances,
- a model that enables organisations to develop collaborative relationships either for one-off projects (project-specific) or as long-term associations (strategic partnering), and
- a process that is formalised within a relationship that might be defined within a charter or a contractual agreement.

For contractors, the continuity of working repeatedly for the same clients is thought to provide a number of benefits for contracting organisations. Clients should normally select their partners from competitive bids based on carefully set criteria aimed at getting best value for money. This initial competition should have an open and known pre-qualification system for bidders.

ALLIANCING

The terms alliancing and partnering do not have the same legal connotations as partnership or joint venture. In project partnering one supplier may sink or swim without necessarily affecting the business position of the other suppliers, unlike alliancing. Therefore, given the operational criteria of an alliance it is vitally important that members of the alliance are selected against rigorous criteria.

PRIME CONTRACTING

A prime contractor is defined as an entity that has the complete responsibility for the delivery and, in some cases, the operation of a built asset and may be either a contractor, in the generally accepted meaning of the term, or a firm of consultants. The prime contractor needs to be an organisation with the ability to bring together all of the parties in the supply chain necessary to meet the client's requirements. There is nothing to prevent a designer, facilities manager, financier or other organisation from acting as a prime contractor. However, by their nature, prime contractors tend to be organisations that have access to an integrated supply chain with substantial resources and skills such as project management.

By establishing long-term relationships with supply chain members it is believed that the performance of built assets will be improved through:

- the establishment of improved and more collaborative ways of working together to optimise the construction process, and
- exploiting the latest innovations and expertise.

The prime contractor's responsibilities might include the following:

- overall planning, programming and progressing of the work,
- overall management of the work, including risk management,
- design co-ordination, configuration control and overall system engineering and testing,
- pricing, placing and administration of suitable subcontractors, and
- systems integration and delivering the overall requirements.

Frameworks

Framework agreements are being increasingly used to procure goods and services in both the private and public sectors.

A framework agreement is a flexible procurement arrangement between parties, which states that works, services or supplies of a specific nature will be undertaken or provided in accordance with agreed terms and conditions, when selected or 'called off' for a particular need. The maximum duration of a framework under current EC rules is four years and can be used for the procurement of services and works. An important characteristic of framework agreements is that inclusion in a framework is simply a promise and not a guarantee of work. Entering into such a framework, however rigorous and costly the selection process, is not entering into a contract, as contracts will only be offered to the framework contractors, supply chains, consultants or suppliers, as and when a 'call off' is awarded under the agreement.

The framework establishes the terms and conditions that will apply to subsequent contracts but does not create rights and obligations. The major advantages of a framework agreement are seen to be:

- it forms a flexible procurement tool,
- the avoidance of repartition when procuring similar items,
- establishment of long-term relationships and partnerships,
- whenever a specific contract call-off is to be awarded, the public body may simply go to the framework contractor that is offering the best value for money for their particular need, and
- reduction in procurement time / costs for client and industry on specific schemes.

New models of construction procurement

In 2011 the government introduced a New Models of Construction Procurement (NMCP) initiative on a trial basis. The initiative consisted of three strategies:

1. Cost Led Procurement,
2. Integrated Project Insurance, and
3. Two Stage Open Book.

A report by the NMCP Working Group in 2019 concluded that although these strategies are not being widely used they appear to be delivering savings of 6–20 per cent compared with traditional competitive procurement.

The common factor of the NMCP models in that they all include the principles of early supply chain involvement and cost transparency in an attempt to improve increased supply chain innovation and value for money.

EU public procurement / environmental impact assessment

EU public procurement and Brexit

The UK left the EU on 31 January 2020; however, the current advice from the EU Commission is that the public procurement regulations and processes will remain broadly unchanged after Brexit, however, the UK will have the status of a 'third country'. In addition, UK entities may no longer have access to the official journal and the key difference for contracting authorities would be the need to send notices to a new UK e-notification service instead of the EU Publications Office, as public procurement entities have a legal obligation to publish public procurement information. Plans for the new UK e-notification service are currently in progress, and a number of organisations have already expressed interest in providing this service. The UK e-notification service will also include opportunities included on the Official Journal of the EU / Tenders Electronic Daily (OJEU / TED). In the longer term it is generally recognised that the impact of Brexit on public procurement is uncertain.

European public procurement law

For projects within the public sector the project manager must be aware of EU public procurement law in so far as it applies to a project. Procurement in the European public sector involves governments, utilities (i.e. entities operating in the water, energy, transport and postal sectors) and local authorities purchasing goods, services and works over a wide range of market sectors, of which construction is a major part. For the purposes of legislation, public bodies are divided into three classes:

1. Central government and related bodies, e.g. NHS Trusts.
2. Other public bodies, e.g. local authorities, universities etc.
3. Public utilities, e.g. water, electricity, gas, rail.

The Directives – theory and practice

The European public procurement regulatory framework was established by the Public Procurement Directives 93/36/EEC, 93/37/EEC and 92/50/EEC for supplies, works and services, and Directive 93/38/EEC for utilities, which, together with the general principles enshrined in the Treaty of Rome (1957), established the principles for cross-border trading (references apply to the Treaty of Rome).

Enforcement Directives (89/665EEC and 92/13EEC) were added in 1991 in order to deal with breaches and infringements of the system by member states.

The Directives lay down thresholds above which it is mandatory to announce the contract particulars. The Official Journal is the required medium for contract announcements and is published five times each week, containing up to 1000 notices covering a wide range of contracts. Major private sector companies also increasingly use the Official Journal for market research. The current thresholds (effective from 1 January 2020 to 31 December 2021) for announcements in the Official Journal are:

- Works: £4,733,252
- Services: £378,660

Note: Figures exclude VAT.

The Directorate-General for Internal Market, Industry, Entrepreneurship and SMEs actively encourages contracting authorities and entities to announce contracts that are below threshold limits. Information on these impending tenders is published by the European Commission in the Official Journal of the European Communities, often otherwise known as the OJEU that is available free of charge, electronically at https://europa.eu.int/. The Directive also clarifies existing law in areas such as the selection of tenderers and the award of contracts, bringing the law as stated into line with judgments of the European Court of Justice.

The EU procurement procedure

AWARD PROCEDURES

The project manager must decide at an early stage which award procedure is to be adopted. The following general criteria apply:

- The minimum number of bidders must be five for the restricted procedure and three for the negotiated and competitive dialogue procedures.
- Contract award is made on the basis of lowest price or most economically advantageous tender (MEAT). Note that from April 2014 MEAT may also now include the 'best price-quality ratio' assessed on the basis of qualitative, environmental and / or social aspects linked to the subject matter of the contract.
- Contract notices or contract documents must provide the relative weighting given to each criterion used to judge the most economically advantageous tender and where this is not possible, award criteria must be stated in descending order of importance.
- MEAT award criteria may now include environmental characteristics (e.g. energy savings, disposal costs) provided these are linked to the subject matter of the contract.

For the second time since their introduction, the EU Public Procurement Directives have recently been amended, the first time being in 2004. According to the EU Commission, the reasons for the changes, which were effective from 17 April 2014, were:

- greater flexibility,
- simplification,
- easier access for SMEs, and
- encouragement for contracting authorities and bidders to interact throughout the procurement process.

However, there still remains a good deal of scepticism amongst those who use public procurement that the changes will make the process easier to navigate.

The new EU procurement regime comprises three new directives:

1. Directive on public procurement, which repeals Directive 2004/18/EC on public works, supply and service contracts.
2. Directive on procurement by entities operating in the water, energy, transport and postal services sectors, which repeals Directive 2004/17/EC.
3. New directive on the award of concession contracts.

The key changes in the process are:

- The introduction of a new procurement regimes for concession awards. Services concessions fall outside the scope of the existing 2004 legislation. Common examples of concessions include running catering establishments in publicly owned sports and leisure facilities, provision of car parking facilities and services, or the operation of toll roads, etc.
- New award procedures, giving scope for more negotiation between contracting authorities and bidders.
- An extension of the grounds for disqualification of bidders.
- Changes to the award criteria.
- New provisions on the modification of contracts post-award.

The choices are as follows;

- **Open procedure** – allows all interested parties to submit tenders.
- **Restricted procedure** – initially operates as the open procedure but then the contracting authority only invites certain contractors, based

on their standing and technical competence, to submit a tender. Under certain circumstances, for example extreme urgency, this procedure may be accelerated.

- **Negotiated procedure** – the contracting authority negotiates directly with the contractor of its choice. Used in cases where it is strictly necessary to cope with unforeseeable circumstances, such as earthquake or flood. Most commonly used in public–private partnership (PPP) models in the UK. From 2014 this may now be used without prior notification.
- **Competitive dialogue** – the introduction of this procedure addresses the need to grant, in the opinion of the European Commission, contracting authorities more flexibility to negotiate on PPP projects. Some contracting authorities have complained that the existing procurement rules are too inflexible to allow a fully effective tendering process. Undoubtedly, the degree of concern has depended largely on how a contracting authority has interpreted the procurement rules as there are numerous examples of PPP projects which have been successfully tendered since the introduction of the public procurement rules using the negotiated procedure. However, the European Commission recognised the concerns being expressed, not only in the UK but also across Europe, and it has sought to introduce a new procedure which will accommodate these concerns. In essence, the new competitive dialogue procedure permits a contracting authority to discuss bidders' proposed solutions with them before preparing revised specifications for the project and going out to bidders asking for modified or upgraded solutions. This process can be undertaken repeatedly until the authority is satisfied with the specifications that have been developed. Some contracting authorities are pleased that there is to be more flexibility to negotiations, however for bidders this reform does undoubtedly mean that tendering processes could become longer and more complex. This in turn would lead to more expense for bidders and could pose a threat to new entrants to the PPP market as well as existing players. According to the Commission's Directorate-General for Internal Market, Industry, Entrepreneurship and SMEs, the introduction of this procedure will enable:
    - dialogue with selected suppliers to identify and define solutions to meet the needs of the procuring body, and
    - awards to be made only on the basis of the most economically advantageous basis.

In addition:

- All candidates and tenderers must be treated equally and commercial confidentiality must be maintained unless the candidate agrees that information may be passed onto others.
- Dialogue may be conducted in successive stages. Those unable to meet the need or provide value for money, as measured against the published award criteria, may drop out or be dropped, although this must be conveyed to all tenderers at the outset.
- Final tenders are invited from those remaining on the basis of the identified solution or solutions.
- Clarification of bids can occur pre- and post-assessment provided this does not distort competition.

To summarise therefore, the competitive dialogue procedure is, according to the Commission, to be used in cases where it is difficult to assess what would be the best technical, legal or financial solution because of the market for such a scheme or the project being particularly complex. However, the competitive dialogue procedure leaves many practical questions over its implementation, for example:

- The exceptional nature of the competitive dialogue and its hierarchy with other award procedures.
- The discretion of the contracting authorities to initiate the procedure – who is to determine the nature of a particular complex project?
- The response of the private sector, with particular reference to the high bid costs.
- The overall value for money.
- The degree of competition achieved as there is great potential for post-contract negotiations.

Note that from April 2014 the competitive dialogue procedure is no longer restricted to complex projects.

In addition to the above procedures the following new procedures have been introduced.

COMPETITIVE PROCEDURE WITH NEGOTIATION

Like competitive dialogue, and actually like the existing negotiated procedure, this is a competitive process where negotiations are to be carried out with all the bidders still in the procurement. The major change from the current negotiated procedure will be that following negotiation on submitted tenders there will be a formal end to the negotiating, and bidders will then be invited to submit a revised tender (very much like the tender phase in competitive dialogue). Another aspect is that it specifies the extent to which the authority can change its requirements during the process. The Directive specifically precludes an authority from making changes to:

- the description of the procurement, and
- the part of the technical specifications which define the minimum requirements of the award criteria.

However, it acknowledges the right to make changes to other parts of the specification provided bidders are given sufficient time to make an adequate response.

Other points to note include:

- As with competitive dialogue, there will be specific grounds which permit its use; these will include that '*due to specific circumstances related to the nature or the complexity of the works, supplies or services or the risks attaching thereto, the contract cannot be awarded without prior negotiations*'.
- The minimum number of bidders to be invited is three.
- It will be possible to hold the negotiation in stages and reduce the number of bidders at the end of a stage.
- The ability to hold an accelerated procedure, currently limited to the restricted procedure, will be extended to the new procedure making it possible to use it in cases of urgency.
- A bidder's solution or other confidential information is not to be revealed to other bidders without specific consent.

The new procedure has much in common with competitive dialogue. What will distinguish it is that, in competitive dialogue the first phase solutions are developed until the authority considers that it has identified one or more capable of meeting its needs and then seeks to formalise positions in a tender; whereas in the new competitive procedure with negotiation, tenders are submitted initially, are then subject to negotiation and then resubmitted to finalise positions.

INNOVATIVE PARTNERSHIPS

This is for use in cases where solutions are not already available on the market.

CONCESSIONS

The new EU concessions regime sets out a basic framework for the award of works and services concessions in the public and utilities sector, subject to certain exemptions in respect of water (such as the disposal or treatment of sewage) with a value of €5.186 million or greater. The new regime leaves the choice of the most appropriate procedure for the award of concessions to individual contracting entities, subject to basic procedural guarantees, including:

- the publication of the 'concession notice' in the Official Journal of the EU advertising the opportunity,
- certain minimum time limits for the receipt of applications and tenders,
- the selection criteria must relate exclusively to the technical, financial and economic capacity of operators,
- the award criteria must be objective and linked to the subject matter of the concession, and
- acceptable modifications to concessions contracts during their term, in particular where changes are required as a result of unforeseen circumstances.

The OJEU announcement procedure involves three stages:

1. Prior information notices (PIN) or indicative notices.
2. Contract notices.
3. Contract award notices (CANs).

Examples of these notices can be found in Annex IV of the Directive.

- **Prior information notice**, or PIN – not mandatory, this is an indication of the essential characteristics of a works contract and the estimated value. It should be confined to a brief statement, and posted as soon as planning permission has been granted. The aim is to enable contractors to schedule their work better and allow contractors from other member states the time to compete on an equal footing.
- **Contract notices** – these are mandatory and must include the award criteria, which can be based on either the lowest price or the most economically advantageous tender, specifying the factors that will be taken into consideration.

- **Contract award notices** – inform contractors about the outcome of the procedure. If the lowest price was the standard criterion, this is not difficult to apply. If, however, the award was based on the 'most economically advantageous tender', then further clarification is required to explain the criteria, e.g. price, period for completion, running costs, profitability and technical merit, listed in descending order of importance. Once established, the criteria should be stated in the contract notices or contract documents.

Electronic tendering

Electronic auctions

The project manager should investigate the possibility of using electronic tendering, as this approach is claimed to offer substantial savings and efficiencies over the traditional approach.

The Internet is making the use of electronic auctions increasingly more attractive as a means of obtaining bids in both public and private sectors; indeed it can be one of the most transparent methods of procurement. At present electronic auctions can be used in both open and restricted framework procedures. The system works as follows:

- The framework (i.e. of the selected bidders) is drawn up.
- The specification is prepared.
- The public entity then establishes the lowest price award criterion, for example with a benchmark price as a starting point for bidding.
- Reverse bidding on a price then takes place, with framework organisations agreeing to bid openly against the benchmark price.
- Prices / bids are posted up to a stated deadline.
- All bidders see the final price.

Technical specifications

At the heart of all domestic procurement practice is compliance with the technical requirements of the contract documentation in order to produce a completed project that performs to the standards of the brief. The project must comply with national standards of being compatible with existing systems and technical performance. References should be made to:

- A standard – a technical specification approved by a recognised standardising body for repeated and continuous application.

- A European Standard – a standard approved by the European Committee for Standardisation (CEN).
- European Technical Approval – a favourable technical assessment of the fitness for use of a product, issued by an approval body designated for the purpose (sector-specific information regarding European technical approval for building products is provided in Directive 89/106/EEC).
- Common technical specification – a technical specification laid down to ensure uniform application in all member states, which has been published in the Official Journal.
- Essential requirements – requirements regarding safety, health and certain other aspects in the general interests that the construction works must meet.

Public procurement beyond Europe

There are no multilateral rules governing public procurement. As a result, governments are able to maintain procurement policies and practices that are trade distortive. That many governments wish to do so is understandable; government purchasing is used by many as a means of pursuing important policy objectives that have little to do with economics – social and industrial policy objectives rank high amongst these. The plurilateral Government Procurement Agreement (GPA) partially fills the void. GPA is based on the GATT provisions negotiated during the 1970s, and is reviewed and refined at meetings (or rounds) by ministers at regular intervals, the last being 2012. Its main objective is to open up international procurement markets by applying the obligations of non-discrimination and transparency to the tendering procedures of government entities. It has been estimated that market opportunities for public procurement increased ten-fold as a result of the GPA. The GPA's approach follows that of the European rules.

Environmental impact assessment (EIA)

About the EIA Directive

The Environmental Impact Assessment (EIA) Directive of 1985 has been amended three times. Directive 2011/92/EU of the European Parliament and the Council of 13 December 2011 on the assessment of the effects of certain public and private projects on the environment, as amended, known as the EIA Directive, requires that an environmental assessment to be carried out by the competent national authority for certain projects which are likely to have significant effects on the environment by virtue, inter alia, of their nature, size or location, before development consent is given. The projects may be proposed by a public or private person.

An assessment is obligatory for projects listed in Annex I of the Directive, which are considered as having significant effects on the environment. These projects include, for example:

- long-distance railway lines,
- airports with a basic runaway length of 2100 m or more,
- motorways, express roads,
- roads of four lanes or more (of at least 10 km),
- waste disposal installations for hazardous waste,
- waste disposal installations for non-hazardous waste (with a capacity of more than 100 tonnes per day), and
- waste water treatment plants (with a capacity exceeding 150,000 population equivalent).

Other projects, listed in Annex II of the Directive, are not automatically assessed. Member states can decide to subject them to an environmental impact assessment on a case-by-case basis or according to thresholds or criteria (for example size), location (sensitive ecological areas in particular) and potential impact (surface affected, duration). The process of determining whether an environmental impact assessment is required for a project listed in Annex II is called screening 2. This particularly concerns for example the following projects:

- construction of railways and roads not included in Annex I,
- waste disposal installations and water treatment plants not included in Annex I,
- urban development projects,
- inland waterways, canalisation and flood-relief works, and
- changes or extensions of Annex I and II projects that may have adverse environmental effects.

Public–private partnerships (PPPs)

One of the basic premises of public–private partnerships (PPPs), is that the private sector is better able to manage public sector facilities because of its superior management expertise and experience. One of the most widely used

PPP models was the private finance initiative (PFI). In PPP arrangements, private sector contractors become the long-term providers of services rather than simply upfront asset builders, combining some or all of the responsibilities for the design, construction, finance (which may be a mixture of public and private sources), facilities management and service delivery of a public service facility.

Over its life span PFI and its replacement PF2, launched in 2012, were heavily criticised by a wide range of organisations. Following the collapse of Carillion in 2018 it was announced that PFI and PF2 contracts would no longer be used. Among the key reasons for Carillon's failure in 2018 was a £375m loss on three PPP contracts, including the construction of Royal Liverpool University Hospital and Metropolitan Midland Hospital. Existing PFI and PF2 contracts will be honoured but no new ones will be signed. There are around 700 active PFI and PF2 deals, with the government estimating that utility payments will cost £199 billion by the 2040s.

The House of Commons public administration and constitutional affairs committee found there are fundamental flaws in the way the government awards contracts because of '*an aggressive approach to risk transfer*'. However, it is widely thought that other models will be developed that permit the private sector to manage and finance public sector projects and these will emerge over time. Other PPP models currently in used in the UK construction sector are listed below (see also Figure 2.11).

- Scottish Futures Trust (SFT) – in 2019/20 SFT had an income of £10.6 million, the majority of which came from the Scottish government and concentrates on infrastructure projects such as roads, hospitals and affordable homes, working collaboratively with public sector clients and industry.
- Frameworks – a framework agreement is a flexible procurement arrangement between parties, which states that works, services or supplies of a specific nature will be undertaken or provided in accordance with agreed terms and conditions, when selected or 'called off' for a particular need. It involves pre-qualification and generally lasts for four or five years.
- ProCure22 (P22) – a construction procurement framework administrated by the Department of Health and Social Care for the delivery of NHS capital projects through public / private co-operation between NHS clients and industry supply chains. Some of P22's principles include:

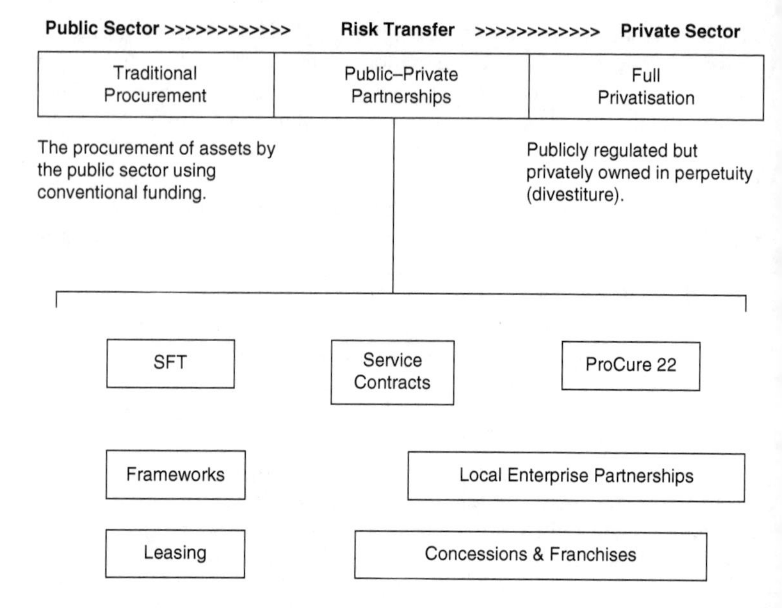

Figure 2.11 PPP models

    - a commitment to partnering,
    - open, honest and transparent relationships with clear accountability, and
    - adherence to good project management through the client.
- Local Partnerships – formed in 2009, Local Partnerships is jointly owned by HM Treasury and the Local Government Association and the Welsh government. Local Partnerships operate in a mainly support role over a wide variety of fields such as housing and waste collection and disposal.
- Concessions – rights conferred on a private company by a government to operate or provide a specific business within the government's jurisdiction for a set period under specific conditions. An example of a concession is the one that exists for the operation of the Channel Tunnel between England and France.

Sustainable procurement

To understand what sustainable procurement means it's important to first understand what is meant by 'sustainable development' and 'procurement'. Sustainable development is a process which enables people to realise their potential and improve their quality of life, now and in the future, whilst protecting the environment. Sustainable development policy should include long-term planning, consideration of impacts beyond the local area (regional, national and international impacts) and the integration of social, economic and environmental issues. Procurement is the whole process of acquisition from third parties covering goods, services and capital projects. The process spans the whole life cycle from initial concept through to the end of the useful life of the asset (including disposal) or end of the services contract.

Sustainable procurement is a key method for delivering an organisation's sustainable development priorities. It is all about taking social and environmental factors into consideration alongside financial factors in making these decisions. It involves looking beyond the traditional economic parameters and making decisions based on the whole life cycle cost, the associated risks, measures of success and implications for society and the environment. Making decisions in this way requires setting procurement into the broader strategic context including value for money, performance management, corporate and community priorities.

At the design stage the project manager needs to be aware of the drivers for sustainability and the impact these have on capital and life cycle costs, as well as the technical requirements of sustainable buildings, so that these are developed into realistic costs and not arbitrary percentage additions. When the project manager is required at this stage to liaise with the client and professional team to determine the client's initial requirements and to develop the client's brief, consideration should be given to the client's overall business objectives, particularly any corporate responsibility targets likely to affect the project. In advising the client on demolition and enabling works, the project manager is advised to consider carrying out a pre-demolition audit to maximise material reclamation and reuse and minimise waste to landfill. The procurement of demolition and enabling works could include evaluation criteria that consider a company's sustainability credentials. Specialists would be required to contribute to meeting the client's objectives and the project targets in the key sustainability areas.

Procurement paths and risk

Single-stage

With the traditional path there is a fair balance of risk between parties. The contractor owns the financial risk of the building works including the performance of the subcontractors, although any alterations to the traditional process will increase risk to the client.

Design and build

With single-stage design and build the contractor assumes the risk for the design development and construction of the project.

Management contracting

With management routes generally, risk lies mainly with the client. With management contracting there is an ongoing sharing of the risks and the client retains certain risks on the subcontractor.

In construction management the balance of risk is more with the client so in-house expertise is essential to reduce risk and make timely decisions.

COST ADVICE

Even though the project is at an early stage in terms of design development the project manager must be able to give the client a reliable estimate of cost. Pre-tender estimating is carried out by the quantity surveyor / cost engineer and cost data may be presented in a number of different formats, outlined below.

Gross external area (GEA)

This approach to measurement is recommended for:

- building cost estimation for calculating building costs for residential property for insurance purposes,
- town planning applications and approvals, and
- rating and council tax.

Note that measurements are taken over areas occupied by internal walls, partitions, columns and attached piers, etc.

Gross internal floor area (GIFA)

Gross internal floor area is the method used most often by quantity surveyors when giving early cost estimates. It is also one of the approaches suggested by NRM 1 when asked to provide cost estimates and is recommended for:

- building cost estimation,
- marketing and valuation of industrial buildings, warehouses, department stores,
- valuation of new homes, and
- apportionment of services charges in property management.

Net internal area (NIA)

This approach to measurement is recommended for:

- marketing and valuation of shops, supermarkets and offices,
- rating shops, and
- apportionment of services charges in property management.

This approach is widely and almost exclusively used by surveyors when determining rents or negotiating leases.

The above definitions are broadly in line with the RICS Code for Measurement Practice (6th Edition). However, in 2015 the RICS introduced a new basis for calculating areas, the International Property Measurement Standards (IPMS).

In 2009 the RICS highlighted a lack of common classifications and a complete absence of standards and transparency in the measurement of buildings, for example office buildings are measured differently nearly everywhere in the world. Property measurements can vary by as much as 24 per cent (Jones Lang LaSalle) depending on the basis of measurement adopted. The driver behind the IPMS is to provide, for the first time, a worldwide principle-based standard which sets out how to measure property assets. Currently standards have been produced for the measurement of:

- office buildings,
- residential buildings,
- retail buildings, and
- industrial buildings.

IPMS can be accessed at https://ipmsc.org/.

The measurements can be used for asset management, benchmarking, construction, facilities management, marketing, property financing, research, transaction, valuation and other purposes. It has to be said that the IPMS has not received widespread acceptance within the professions.

The RICS New Rules of Measurement 1: Order of Cost Estimate and Cost Planning for Capital Building Works

NRM 1 Order of Cost Estimating and Elemental Cost Planning for Capital Building Works was launched in March 2009, with a second edition following in April 2012. It aims to provide a comprehensive guide to good cost management of construction projects.

The rationale for the introduction of the NRM 1 is that it provides:

- a standard set of measurement rules that are understandable by all those involved in a construction project, including the employer, thereby aiding communication between the project / design team and the employer,
- direction on how to describe and deal with cost allowances not reflected in measurable building work, and
- a more universal approach than the SMM7, which was considered UK-centric.

The NRM has been developed to:

- modernise the existing standards that many of those involved in measuring building work have been used to working with,
- improve the way that measurement for cost planning and bills of quantities has been delivered, and
- begin addressing a common standard for life cycle cost planning and procurement of capital building works, and the life cycle of replacement and maintenance works.

It is important to understand that the NRM 1 is a toolkit for cost management, not just a set of rules for how to quantify building work. As a toolkit, the NRM 1 provides guidance on:

- how measurement changes as the design progresses – from high level cost / m^2 or cost / functional units to more detailed measurement breakdowns of elements and sub-elements,

- total project costs – it provides guidance on how all cost centres can be considered and collated in to the project cost plan,
- risk allowances based on a properly considered assessment of the cost of dealing with risks should they materialise – dispensing with the use of the widely mismanaged concept of contingency,
- total project fees – it provides guidance on how fee and survey budgets can be calculated,
- the suggested design and survey information that a client needs at each RIBA Stage / OGC Gateway for the quantity surveyor to be able to provide more certainty around their cost advice,
- the suggested key decisions that clients need to make at each RIBA Stage / OGC Gateway, and
- a framework for codifying cost plans so they can be converted into works packages for procurement and cost management during construction.

The Building Cost Information Service Standard Form of Cost Analysis (SFCA) was first produced in 1961 when the bill of quantities was king, and was subsequently revised in 1969 and 2008, and has been the industry norm for the last forty years. In April 2012, to coincide with the publication of NRM 1, the SFCA was also updated so that now both the SFCA and NRM 1 are in the same format.

The RICS formal cost estimation and cost planning stages align with the RIBA Plan of Work and OGC Gateways. One of the factors that has driven NRM 1 is the lack of specific advice on the measurement of building works solely for the purpose of preparing cost estimates and cost plans. As someone who has tried to teach cost planning and estimating for the last forty years, I am acutely aware that students, as well as practitioners, are often confused as to how estimates and cost plans should be prepared, resulting in the process taking on the air of a black art! This situation has led to an inconsistent approach, varying from practice to practice, leaving clients a little confused. It is also thought that the lack of importance of measurement has been reflected in the curriculum of degree courses, resulting in graduates unable to measure or build up rates, a comment not unknown during the last fifty years or so.

As illustrated in Figure 2.12 the process of producing a cost estimate and cost planning has been mapped against the RIBA plan of work and OGC Gateway process. It shows that the preparation and giving of cost advice is a continuous process that in an ideal world becomes more detailed as the information flow becomes more detailed. In practice it is likely that the various stages will merge and that such a clear-cut process will be difficult to achieve.

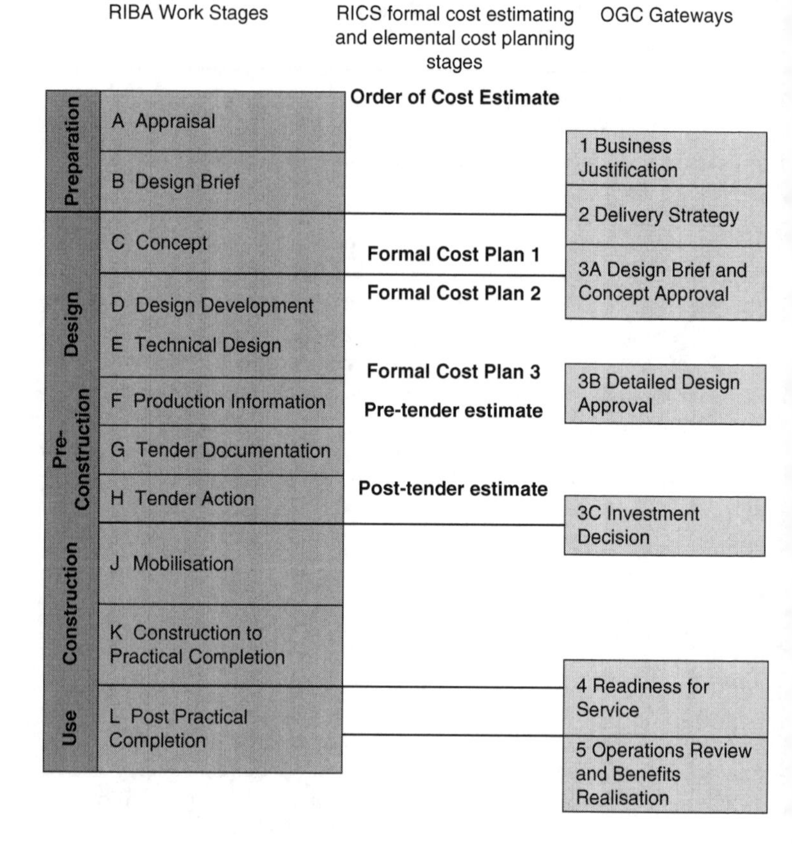

Figure 2.12 RIBA Plan of Work and OGC Gateway compared

NRM 1 suggests that the provision of cost advice is an iterative process that follows the information flow from the design team as follows:

- Order of cost estimate
- Formal cost plan 1
- Formal cost plan 2
- Formal cost plan 3
- Pre-tender estimate.

There would therefore appear to be two distinct stages in the preparation of initial and detailed cost advice:

1. **Estimate** – an evolving estimate of known factors. Is the project affordable? The accuracy at this stage is dependent on the quality of the information. Lack of detail should attract a qualification on the resulting figures. At this stage information is presented to the client.
2. **Cost plan** – a critical breakdown of the cost limit for the building into cost targets for each element. At this stage it should be possible to give a detailed breakdown of cost allocation.

In addition the NRM 1 approach divides cost estimates and cost plans into five principal cost centres;

1. works cost estimate,
2. project / design team fees estimate,
3. other development / project cost estimate,
4. risk allowance estimate, and
5. inflation estimate.

The Order of Cost Estimate and Cost Plan Stages have differing recommended formats. Compared to the BCIS (SFCA) the NRM 1 format does provide a greater range of cost information to the client covering the following building works, including facilitating works:

- main contractor's preliminaries,
- main contractor's profit and overheads,
- project / design team fees,
- other development / project costs,
- risk,
- inflation,
- capital allowances, land remediation relief and grants, and
- VAT assessment.

RICS New Rules of Measurement – order of cost estimate format

A feature of NRM 1 is the detailed lists of information that is required to be produced by all parties to the process; the employer, the architect, the mechanical and electrical services engineers and the structural engineer all have substantial lists of information to provide. There is an admission that the accuracy of an order of cost estimate is dependent on the quality of the information supplied to the quantity surveyor. The more information provided, the more reliable the outcome will be, and in cases where little or no information is provided, the quantity surveyor will need to qualify the order of cost estimate accordingly.

The development of the estimate / cost plan starts with the order of cost estimate.

INITIAL RISK REGISTER: RISK ALLOWANCE ESTIMATE

Risk is defined as *the amount added to the base cost estimate for items that cannot be precisely predicted to arrive at the cost limit.*

The inclusion of a risk allowance in an estimate is nothing new; what perhaps is new, however, is the transparency with which it is dealt with in NRM 1. It is hoped, therefore, that the generic cover-all term 'contingencies' will be phased out. Clients have traditionally homed into contingency allowances wanting to know what the sum is for and how it has been calculated. The rate allowance is not a standard percentage and will vary according to the perceived risk of the project. Just how happy quantity surveyors will be to be so up front about how much has been included for unforeseen circumstances or risk will have to be seen. It has always been regarded by many in the profession that carefully concealing pockets of money within an estimate for extras / additional expenditure is a core skill.

So how should risk be assessed at the early stages in the project? It is possible that a formal risk assessment should take place, and this would be a good thing, using some sort of risk register. Obviously, the impact of risk should be revisited on a regular basis as the detail becomes more apparent.

Risks are required to be included under the following four headings:

- **Design development risks**, which may include such items as:
  - inadequate or unclear project brief,
  - unclear design team responsibilities,
  - unrealistic design programme,
  - ineffective quality control procedures,
  - inadequate site investigation,
  - planning constraints / requirements, and
  - soundness of design data.

- **Construction risks**, for example:
  - inadequate site investigation,
  - archaeological remains,
  - underground obstructions,
  - contaminated ground,
  - adjacent structures (i.e. requiring special precautions),
  - geotechnical problems (e.g. mining and subsidence),
  - ground water,
  - asbestos and other hazardous materials, and
  - invasive plant growth.
- **Employer's change risk**, for example:
  - specific changes in requirements (i.e. in scope of works or project brief during design, pre-construction and construction stages),
  - changes in quality (i.e. specification of materials and workmanship),
  - changes in time,
  - employer driven changes / variations introduced during the construction stage,
  - effect on construction duration (i.e. impact on date for completion), and
  - cumulative effect of numerous changes.
- **Employer's other risks** – this section has a long list of items including for example:
  - Project brief:
    - end user requirements,
    - inadequate or unclear project brief, and
    - employer's specific requirements (e.g. functional standards, site or establishment rules and regulations, and standing orders).
  - Timescales:
    - unrealistic design and construction programmes,
    - unrealistic tender period(s),
    - insufficient time allowed for tender evaluation,
    - contractual claims,
    - effects of phased completion requirements (e.g. sectional completion),
    - acceleration of construction works,
    - effects of early handover requirements (e.g. requesting partial possession),
    - postponement of pre-construction services or construction works, and
    - timescales for decision-making.

- Financial:
  - availability of funds,
  - unavailability of grants / grant refusal,
  - cash flow effects on timing,
  - existing liabilities (i.e. liquidated damages or premiums on other contracts due to late provision of accommodation),
  - changing inflation,
  - changing interest rates,
  - changing exchange rates, and
  - incomplete design before construction commences.
- Management:
  - unclear project organisation and management,
  - competence of project / design team, and
  - unclear definition of project / team responsibilities.
- Third party:
  - requirements relating to planning (e.g. public enquiries, listed building consent and conservation area consent),
  - opposition by local councillor(s),
  - planning refusal,
  - legal agreements, and
  - works arising out of party wall agreements.
- Other:
  - insistence on use of local work people,
  - availability of labour, materials and plant,
  - statutory requirements,
  - market conditions,
  - political change,
  - legislation, and
  - force majeure.

3

Construction / RIBA Plan of Work Stage 5

For the project manager the priorities during the construction phase are:

- to ensure the resolution of design queries as they arise in order that the client's project is delivered to the required standards, and
- to put systems in place to monitor the progress and cost of the works, as during the construction and site operations stages there is the potential for costs to escalate and programmed dates to slip.

It is claimed that the increasing use of BIM should mean delays, redesign and the associated costs should be reduced during construction as design clash problems between items such as service installations and structure should be reduced or eliminated.

Figure 3.1 illustrates the various stages in the construction and post-construction phases (Chapter 4) of a project based on JCT (16).

ROLES FOR THE PROJECT TEAM

It is important for the project manager to understand the roles and responsibilities of the various project team members during the construction phase in order for project to progress smoothly. These can be summarised as follows.

Client

The client has a duty to:

- satisfy themselves that the project is being constructed in accordance with the brief,

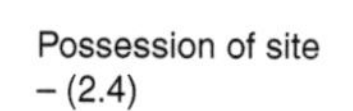

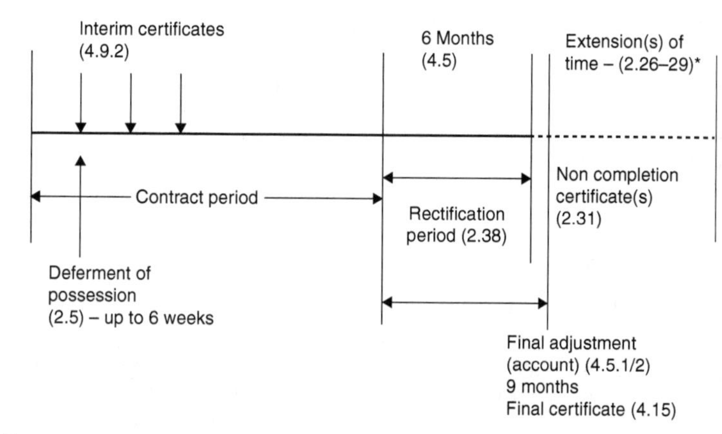

* Note: Extension(s) of time **do not** give a contractor an automatic right to a claim for loss and expense

Figure 3.1 Contract administration

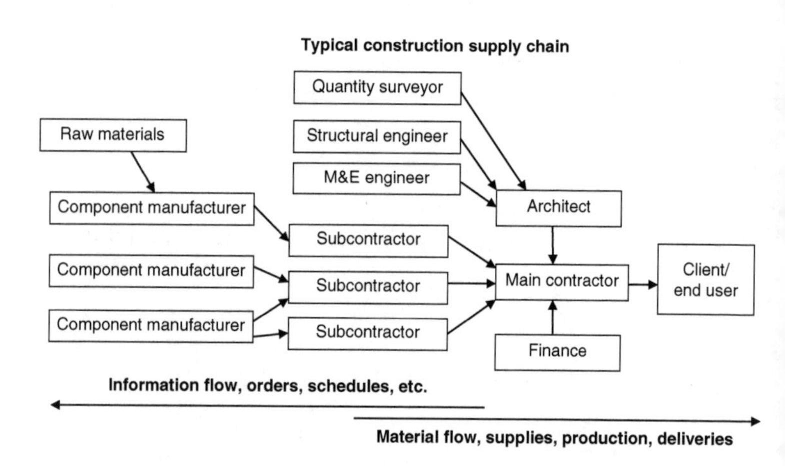

Figure 3.2 Construction supply chain

- ensure that sufficient funds are available to honour interim and final payments, and
- make any decisions regarding change orders / variations in a timely manner.

Architect

For the architect the construction phase involves:

- site visits at agreed on intervals to evaluate the work,
- reports of deviations from the contract documents,
- communication between the owner and the contractor regarding matters related to contract documents,
- rejection of work not conforming to the contract documents,
- certification of payments to the contractor,
- review and approval (or appropriate action) of contractor's submittals such as drawings, product data and samples,
- preparation of change orders and construction change directives,
- review of changes in work including contract sum or contract time and recommendations to owner, and
- inspections to determine the date / dates of substantial completion and final completion.

Clerk of works

The clerk of works is the architect's inspector on-site on a day-to day basis. He or she is the person to whom the contractor will turn when problems arise. The clerk of works is able, under JCT (16), to issue instruction to the contractor but these must be confirmed in writing within two working days, although in practice this is seldom the case.

Structural engineer

The structural engineer works with the architect to provide and approve details for the contractor and subcontractors. In addition the structural engineer will be available to solve site queries and inspect completed works to ensure compliance with the specification and design parameters.

Quantity surveyor

It is the responsibility of the quantity surveyor to ensure effective cost management during the construction phase. When the contract is based on conventional procurement paths, this will include:

- preparing and issuing financial statements for the client (see Appendix A); the intervals between statements will depend on the size and nature of the project and the client's requirements,
- measuring and valuing the works executed by the main contractor,
- agreeing monthly payments on account with the contractor,
- agreeing the final account with the main contractor, and
- agreeing any contractual claims with the contractor.

Main contractor's team

In the terms of the JCT(16) (2.1) standard form of contract: '*The contractor shall carry out and complete the works in a proper and workmanlike manner and in compliance with the contract documents, the construction phase plan and other statutory requirements*'. To help achieve this the contractor relies on:

- the construction manager,
- the site agent, and
- subcontractors and suppliers.

PROJECT QUALITY MANAGEMENT

It is important for the project manager to be aware that some processes, although traditionally dealt with towards the end of the contract, really should be addressed at an early stage, as the earlier their introduction, the greater the potential for efficiency savings, for example:

- post occupancy evaluation,
- Soft Landings, and
- environmental management systems (EMS).

Processes available to the project manager at the construction stage include:

- quality assurance,
- regular site inspections,

- review of progress,
- management of meetings,
- project audit (see Chapter 5), and
- environmental management systems (EMS).

Quality audit

Who needs a quality audit?

- The project manager, seeking unbiased and comprehensive information from groups or individuals within the project organisation.
- The organisation, which seeks to identify the errors made and track their causes, and learns not to repeat them.
- The client, who can relate the value of project development to their own actions and decisions.
- Any external stakeholders or sponsors of the project, for example financial institutions, government agencies, consumer groups, environmental or religious organisations and social groups.

The project manager should establish and implement an appropriate process to manage quality management of the project. There are three aspects to take into account:

1. **Quality planning** – identifying which quality standards are relevant to the project, and determining how to satisfy them.
2. **Quality assurance (QA)** – the process of evaluating overall project performance on a regular basis to provide confidence that the project will satisfy the relevant quality standards. QA also refers to the organisational unit that is assigned responsibility for quality assurance.
3. **Quality control (QC)** – the process of monitoring specific project results to determine if they comply with relevant quality standards and identifying ways to eliminate causes of unsatisfactory performance. QC also refers to the organisational unit that is assigned responsibility for quality control.

The various instruments to ensure the defined quality will vary from stage to stage but could include:

- value engineering / management (see Chapter 2),
- pre-qualification questionnaires (see Chapter 5),

- quality audits, and
- benchmarking (see Chapter 5).

OFF-SITE CONSTRUCTION AND MODERN METHODS OF CONSTRUCTION

Off-site construction, often referred to as modern methods of construction (MMC), is being increasingly employed as part of construction projects. Currently there is no definitive definition of MMC and there continues to be considerable debate within the industry as to what constitutes MMC. There are a number of systems which fall within the umbrella of MMC and for the sake of guidance MMC should be considered as:

> *Those systems which provide an efficient product management process to provide more products of better quality in less time. It has been defined in various ways: pre-fabrication, offsite production and offsite manufacture (OSM). But while all OSM is MMC, not all MMC is OSM.*

Types of MMC can include:

- non-off-site manufactured modern methods of construction,
- sub-assemblies and components – off-site manufactured,
- hybrid – off-site manufactured,
- panellised – off-site manufactured, and
- volumetric – off-site manufactured.

The role of off-site manufacture (OSM) on any project is broad ranging and can be extended from the supply of common M&E products such as multi-service corridor modules with pipework, ductwork and electrical containment to much larger products where the building structure and fabric is incorporated, for example in multi-sectional plant rooms.

In reality, the only limitation to the size of product which can be designed and manufactured off-site is the transportation, where vehicle size and highway regulations dictate and commonly restrict load sizes to 14m x 4m x 4m. Though the amount of services that come in modular form varies from project to project, as a rule of thumb, 60–75 per cent of M&E 'first fix' installation works are off-site (that is, the infrastructure for services like cables and pipes for water and electrical services). After that, 70–100 per cent of M&E work in plant rooms is now off-site. There are almost no restrictions on the use of off-site manufacture – the main considerations that can impact a decision to use it include site logistics and access. Early engagement during the design and planning phases will identify these points and enable restrictions to be incorporated. The UK construction turnover is approximately £100 billion per annum, of which around 6 per cent or £6 billion is currently in MMC. There is evidence to suggest that eventually OSM / MMC can deliver up to 50 per cent of construction spend at an annual growth of around 25 per cent per annum, the limiting factor being the current lack of core skills. Project managers can generally work on the basis that off-site products could save as much as 15 per cent compared with the traditional installation, and at worst give a cost neutral outcome.

Some of the direct and indirect cost benefits associated with OSM / MMC are:

- increased quality of construction / quality control,
- increased speed of construction / reduced programme length,
- reduced bad weather delays / extensions to programme,
- reduced disruption on-site,
- reduced site resources,
- funding required for shorter period,
- reduced risk for the client,
- greater predictability of programme and cost,
- improved health and safety record,
- reduction in waste,
- the client's ability to produce prototypes of products which can be inspected and trialled prior to construction works starting on-site,
- possible tax relief on monies invested in R&D,
- a move towards sustainable construction with the emphasis on prefabrication and off-site assembly,
- becoming more innovative to streamline the construction process,
- the opportunity to develop partnering between contractors and suppliers, and
- reduced environmental impact (BREEAM).

When considering the use of OSM / MMC the following points should be considered:

- Pre-construction – in order to gain the maximum advantages, early involvement and consideration of OSM / MMC is essential.
- Procurement route – does the procurement route lend itself to OSM / MMC?

- Design issues – who is responsible for detailed design?
- Interface – between factory-made parts and on-site assembly.
- Tolerance – the tolerances allowed for each system and the overall impact.
- Buildability – how easy or difficult will overall assembly be?

Off-site construction requires skills that are different to those needed for traditional construction. In particular, off-site construction professionals need a greater understanding of the interaction between principles of design, construction, manufacturing and engineering. If the UK construction industry is to exploit the potential of OSM, multi-skilling, collaboration and greater flexibility within job roles are crucial.

Key ingredients to success can be summarised as:

- a client who understands the modular design, manufacture and installation process and is committed to engage in this process,
- a robust project programme co-ordinated by a modular company that can deliver a quality product to timescale and budget,
- a skilled, proactive and integrated core design team, technical and project management team, architects, civil and structural engineers, M&E engineers, etc., and
- planning authority and building control commitment to engage in a fast track process.

The project manager should be aware of the impact of off-site construction on the construction process, as during the past decade or so, the use of off-site or prefabricated units in the construction of both new and refurbished buildings has become commonplace. Perhaps the most common example of this is the use of bathroom / wet room pods and that there is, in some cases, a need for what has been termed interface management. This could involve considering the impact of the use of off-site units at an earlier stage that has previously been the case.

Housebuilder Barratt met its target to build 20 per cent of its homes using an element of MMC a year ahead of its 2020 target, and has set itself a new target of using MMC in 25 per cent of homes by 2025.

ENVIRONMENTAL MANAGEMENT SYSTEMS (EMS)

An increasing number of client organisations and contractors run environmental management systems (EMS). The project manager should ensure that

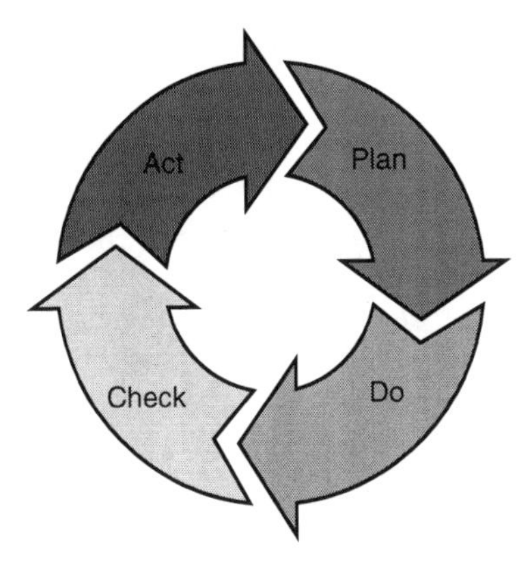

Figure 3.3 The plan–do–check–act (PDCA) cycle

any new project complies with the EMS. An EMS is a set of processes and practices that enable an organisation to reduce its environmental impacts and increase its operating efficiency. The EMS framework is based on the plan–do–check–act (PDCA) cycle (Figure 3.3).

Basic elements of an EMS include:

- reviewing the company's environmental goals,
- analysing its environmental impacts and legal requirements,
- setting environmental objectives and targets to reduce environmental impacts and comply with legal requirements,
- establishing programmes to meet these objectives and targets,
- monitoring and measuring progress in achieving the objectives,
- ensuring employees' environmental awareness and competence, and
- reviewing progress of the EMS and making improvements.

What is an environment management system?

An EMS is a framework that helps a company achieve its environmental goals through consistent control of its operations. The assumption is that this increased control will improve the environmental performance of the company. The EMS itself does not dictate a level of environmental performance that must be achieved; each company's EMS is tailored to the company's business and goals. EMS is voluntary and can be either:

- developed as an in-house EMS, or
- compliant with ISO 14001:2015.

ISO 14001 is in fact a series of international standards on environmental management. It provides a framework for the development of an environmental management system and the supporting audit programme. The ISO 14001 series emerged primarily as a result of the Uruguay round of the GATT negotiations and the Rio Summit on the Environment held in 1992. While GATT concentrates on the need to reduce non-tariff barriers to trade, the Rio Summit generated a commitment to protection of the environment across the world.

After the rapid acceptance of ISO 9000 and the increase of environmental standards around the world, the International Organization for Standardization (ISO) assessed the need for international environmental management standards. They formed the Strategic Advisory Group on the Environment (SAGE) in 1991, to consider whether such standards could serve to:

- promote a common approach to environmental management similar to quality management,
- enhance organisations' ability to attain and measure improvements in environmental performance, and
- facilitate trade and remove trade barriers.

In 1992, SAGE's recommendations created a new committee, TC 207, for international environmental management standards. This committee and its sub-committees included representatives from industry, standards organisations, government and environmental organisations from many countries. What developed was a series of ISO 14000 standards designed to cover:

- environmental management systems,
- environmental auditing,
- environmental performance evaluation,
- environmental labelling,
- life cycle assessment, and
- environmental aspects in product standards.

ISO 14001 was first published as a standard in 1996 and it specifies the actual requirements for an environmental management system. It applies to those environmental aspects over which an organisation has control and where it can be expected to have an influence. ISO 14001 is often seen as the cornerstone standard of the ISO 14000 series. It specifies a framework of control for an environmental management system (EMS) and is the only ISO 14000 standard against which it is currently possible to be certified by an external certification body. However, it does not in itself state specific environmental performance criteria.

This standard is applicable to any organisation that wishes to:

- implement, maintain and improve an EMS,
- assure itself of its conformance with its own stated environmental policy,
- demonstrate conformance,
- ensure compliance with environmental laws and regulations,
- seek certification of its EMS by an authorised external certification body, and
- make a self-determination of conformance.

Other standards in the series are actually guidelines, many to help an organisation achieve registration to ISO 14001. These include the following:

- ISO 14004:2016 provides guidance on development and implementation of an EMS.
- ISO 14010 provides general principles of environmental auditing (now superseded by ISO 19011).
- ISO 14011 provides specific guidance on auditing an environmental management system (also superseded).
- ISO 14012 provides guidance on qualification criteria for auditors and lead auditors (also superseded).
- ISO 14013/5 provides an audit programme review and assessment material.
- ISO 14020+ covers labelling issues.
- ISO 14030+ provides guidance on performance targets and monitoring within an EMS.
- ISO 14040+ covers life cycle issues

WORKS ON-SITE

Meetings

The project manager should schedule a number of regular meetings at the start of the construction phase with key members of the project team.

These meetings could cover the following aspects:

- site meetings,
- project meetings,
- initial meeting,
- ongoing meetings, and
- general review meetings.

Site meetings

Site meetings are usually held every month, although at critical periods in the project this may be shortened to every two weeks. The main objective of these meetings is to review progress and they are usually chaired by the architect / CA. Those present usually include:

- architect and clerk of works,
- project manager,
- quantity surveyor,
- contractor / main subcontractors, and
- engineers – structural / mechanical and electrical.

It is usual during these meetings that actual progress is compared with planned progress and any claims that the contractor may have for extensions of time are recorded along with outstanding variation / change orders. It is vitally important that an accurate record of the proceeding is minuted as they may form the basis for a claim by the contractor at the end of the project. Therefore, it is important that the project manager checks the minutes for accuracy and notes any possible financial and planning implications. Site meetings also usually include, at some point in the proceedings, a visit to the site.

A typical basic agenda for a monthly site meeting is as follows:

- matters arising from previous meetings,
- main contractor's statement on progress (main contractor / subcontractor),
- reports, as appropriate, from the architect / CA, quantity surveyor, etc.,
- report by the contractor of work by statutory undertakings (e.g. electricity, telecoms, gas, etc.),
- outstanding information:
    - drawings may be issued,

- statement of outstanding change orders / variations, and
- approval and consents, planning, building regulations, etc.

Project meetings

Running and organising successful meetings are vital parts of the contract administration process. The agendas for project meetings fall into two formats, the initial meeting and ongoing meetings.

INITIAL MEETING

Matters to be covered could include:

- introduction of relevant representatives,
- communication and site responsibility issues,
- scheduling of project meetings,
- reviewing the contractor's construction programme,
- reviewing payment schedules and processes,
- processes for resolving construction discrepancies,
- reviewing the list of subcontractors and suppliers,
- processes for dealing with uncertain work,
- health and safety matters,
- any other design clarifications or procedural issues, and
- any of the owner's concerns and questions.

ONGOING MEETINGS

The agenda should include:

1. A review of:
   (a) the construction schedule,
   (b) relevant quality issues, and
   (c) costs and overall budget.
2. Construction activity issues.
3. Unanticipated delays.
4. Design clarifications.
5. Contract variations, owner's concerns.
6. Unforeseen extra work, decay, etc.
7. Health and safety matters.

General review meetings

Matters covered should include:

- monitoring and approving work quality,
- approving progress payment claims,
- approving variation or provisional cost claims,
- interpreting contract document requirements,
- providing explanations, interpretations, clarifications and extra instructions,
- updating drawings, etc. if required,
- providing variation instructions (and any role in amending consents),
- reviewing contractor's variation claims,
- verifying substantial completion, and
- verifying final completion.

Ten rules for running productive meetings

1. **Make the objective of the meeting clear** – a meeting should have a specific and defined purpose. Be clear about why you are meeting. Write down the purpose and objectives of the meeting. You should be able to do this in a sentence or two.
2. **Decide who needs to be at the meeting** – invite only those people whose input is necessary. If you're trying to find the solution to a problem, invite the people who will be good sources of information for a solution.
3. **Decide how much time is needed for the meeting and stick to the schedule** – prepare an agenda that lays out everything you plan to cover in the meeting and consider incorporating a timeline that allots a certain number of minutes to each item. Email it to people in advance.
4. **Start on time, end on time** – if you have responsibility for running regular meetings, establish a reputation for being someone who starts and ends promptly. People appreciate it when you understand that their time is valuable. Sixty minutes is generally the longest time people can remain truly engaged.
5. **Give people as much advance notice as possible** – send out an announcement of the meeting by email. In addition to stating the purpose of the meeting, include the date, time, location and how long the meeting will run, including a call-in number if appropriate.
6. **Choose a convenient time** – schedule meetings in core hours and try to avoid meeting very early or very late in the day.

7. **Distribute the agenda and any meeting handouts ahead of time so people can prepare** – send the agenda out at least two days in advance of the meeting. This way, the agenda also serves as a reminder for the meeting.
8. **Send everyone a reminder a few days before the meeting** – this isn't necessary if you've sent out an agenda and reminder together.
9. **Ban technology** – if people are allowed to bring iPads or mobile devices into the room, they won't be focusing on the meeting or contributing to it. Instead, they'll be emailing, surfing the web or just playing around with their technology.
10. **Follow up** – keep minutes and distribute in good time, detailing the responsibilities given, tasks delegated and any assigned deadlines.

CONSTRUCTION

During the construction phase the project manager should pay particular attention to the following:

1. Cost control / financial control – keeping up to date with the current financial status of the project.
2. Acceleration of the contract programme may be required.
3. Insolvency / bankruptcy / sequestration of main or subcontractors.
4. Generally ensuring good supply chain relationships and management.

Cost control / financial reporting

Financial statements

Throughout the project the project manager will be expected to provide the client with an accurate statement of the financial position of the contract works. This statement, usually prepared by the quantity surveyor on a monthly or quarterly basis, takes into account all adjustments to the original cost including such items as:

- adjustments for variations,
- adjustment for provisional sums,
- agreed and anticipated contractor's claims,
- anticipated variation, and
- fluctuations (if applicable).

See Appendix C for an example of a financial statement. Note that the quantity surveyor should be able to provide detailed back-up for all figures included.

Interim payments

During the contract period the contractor usually receives monthly payments on account, the extent of which are determined by the contractor and client's quantity surveyors. It should be noted that inclusion and payment of works in interim payments is not an acceptance of the quality of the work and sums included in interim payments may be subsequently omitted.

Cash flow forecasting

Cash flow forecasting is an essential tool for the project manager to ensure the financial integrity of the project and can be used for a number of purposes including:

- by a client to secure funding,
- by a client to illustrate when and how much is due to the contractor at various stages in the contract period,
- by a contractor to reconcile income with expenditure, and
- by a project manager to compare anticipated progress against actual progress in terms of cash flow.

Cash flow forecasting takes on an extra significance when using NEC contracts, as the programme and activity schedules are specifically referenced within the contract conditions. Therefore, cash flow forecasts that are produced in accordance with these documents can be used by the project manager to assess any compensation events, early warnings or programme revisions before accepting them.

Variations and change orders

The Housing Grants, Construction and Regeneration Act 1996 commonly known as the Housing Act was updated by the Local Democracy, Economic Development and Construction Act 2009 and became effective from November 2011. The Act was introduced, amongst other things, in an attempt to improve payment practice within the construction industry and in particular the practice of 'pay when paid'.

PAYMENTS TO THE SUPPLY CHAIN

At its most basic, the idea is that it will no longer be possible for anyone who owes money down the supply chain to hold up paying it because they themselves have not been paid – the so called 'pay when paid' practice – with the introduction of the concepts of Payment Notices and Pay-Less Notices.

PAYMENT NOTICES

A typical contract will require the employer to issue a Payment Notice not later than five days after the contract payment date is due. This Notice must specify what amount it considers is due to the contractor and the basis on which the amount is calculated. The employer then has to pay that amount. This is called the notified sum. If the employer thinks less is due than was agreed a Pay-Less Notice is served. Such a notice has to specify why that lesser amount is due and give detail information as to how it has been calculated.

WITHHOLDING NOTICES

Withholding Notices are notices saying why payment is being withheld; again full details must be given.

DEFAULT NOTICES

Default Notices can be issued by a contractor when an employer fails to serve a Payment Notice within the specified timescale. The notice must specify what sum the contractor says is due and the basis upon which it is calculated. The employer must pay this sum unless they are entitled to serve a Pay-Less Notice because it does not agree the sum set out in the Default Notice.

PAYEE-LED PAYMENT PROCESS

The Act introduces a new optional 'payee-led' process. Under this procedure the contractor issues the Payment Notice. The employer must pay the notified sum, although it can issues a Pay-Less Notice if it considers the contractor is not due to all that it has claimed. The contractor cannot then serve a Default Notice as it has already indicated what sum it expects to receive.

Note that his payee-led process can only be used if specifically incorporated into the contract.

SUSPENSION OF PERFORMANCE

Under the Act, if a paying party fails to pay what is due, a receiving party can suspend performance of part of their obligations under the contract for non-payment.

It will not be possible to handcuff the Adjudicator's jurisdiction in relation to his own costs. Clauses that previously stated that the party commencing the adjudication process should pay the Adjudicator's costs are outlawed as this meant that almost always the contractor has to pay those costs as it was the party being denied its money.

ADJUDICATION

Adjudication in the case of a dispute will now apply to oral contracts. Adjudicators will be able to award costs on the merits of a case.

IN CONCLUSION

- The Act now applies to all construction contracts whether in writing or not.
- Pay when paid clauses are prohibited. This means that simply because an employer is yet to certify the main contractor, the main contractor can no longer rely on such clauses to prevent paying any subcontractor.
- Clauses in contracts which state that the release of retention is conditional on the issue of a Certificate of Making Good Defects (or similar) are outlawed. Contractors, wherever they are in the chain, should expect to see the release of retention linked to events in their own contract giving them more control over the release of their retention.

Cash flow

For contractors and subcontractors cash flow is the life blood of their business operations. There are various ways by which a contractor / subcontractor receives funds during the contract:

- Although uncommon in the UK, contractors may request an advance of work before starting on-site. If this is done then the project manager should insist that a performance bond is taken out by the contractor to guarantee the sum advanced.
- Through monthly valuations / interim payments – these payments are related to actual progress and productivity and are a realistic snapshot of progress. Monthly valuations are reliable in terms of timing but can be affected by such factors as inclement weather, which may impact on productivity / progress and consequently cash generation.
- Stage payments are usually payments that become pre-agreed milestones and, unlike monthly valuations, are not related to progress. This makes stage payments less than accurate when trying to assess progress against cash generated and as such should be carefully monitored by the project manager.

The classic S-curve profile illustrated in Figure 3.4 illustrates the typical pattern of income flowing from a construction project, with a slow start and finish, and cash flow peeking during the mid-third of the project.

The value of cash flow can either be produced based on the construction contract value, which will be the amount due to the main contractor and will exclude elements such as professional fees, VAT, etc., or the overall

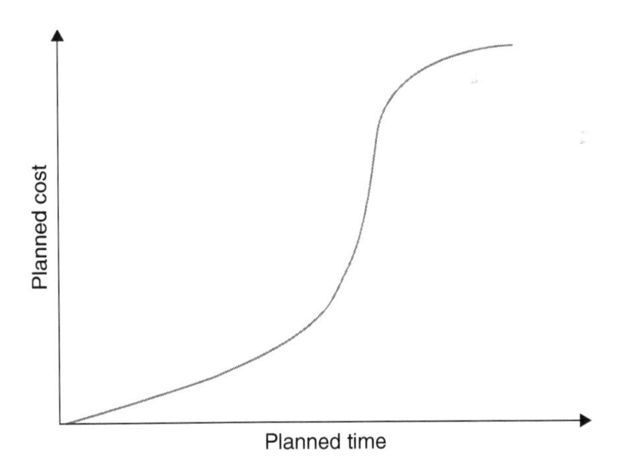

Figure 3.4 S-curve

project value. The project manager should have a discussion with the client to point out the two differences of approach to preparing cash flow forecasts. The cash flow forecast can be compared with actual expenditure and can be an indicator to potential insolvency of contractors / subcontractors if the forecast milestones are not met.

Other items that can distort the forecasting calculation and should be noted are:

- The employer must decide whether the cash flow forecast is to show the valuation date, the certificate date or the actual payment date. Whatever is chosen it must be clearly noted on the cash flow forecast; otherwise a problem could arise if it is assumed that the valuation date was shown when in reality the forecast was showing the payment date.
- The time lag between valuation date and payment date could land the client in an embarrassing financial situation.
- Changes to design, inclement weather, variation orders and labour shortages.
- Delays in agreeing contract claims.
- Unforeseen costs, for example addition substructure works, the discovery of antiquities and ground contamination, etc.
- Strikes or material shortages.

The UK construction industry is notorious for poor payment. It's not uncommon for contractors and subcontractors to have to wait 90–120 days for payment. In 2016 the UK government established a prompt payment code and from 1 September 2019, a supplier who bids for a government contract above £5 million per annum is required to answer questions about their payment performance. The code requires a commitment from contractors to pay 95 per cent of all suppliers within 60 days (see www.promptpaymentcode.org.uk.)

Acceleration

For a number of reasons when work commences on-site there may be a request from either the client or the contractor to accelerate the contract works. Acceleration involves increasing the originally planned or current rate of progress of the works in order to complete the project earlier than would be otherwise be the case.

When considering acceleration the project manager should investigate the following factors:

- Does the contract allow for acceleration?
- Is a separate agreement required?
- How can acceleration be achieved?
- Is the activity to be accelerated on the critical path?
- What costs are involved?
- What costs can be saved by achieving an earlier than planned completion date?
- Is the contractor to guarantee the earlier completion date?
- What happens if early completion is not achieved?

Acceleration can be used when:

- the contractor wishes to complete early to reduce costs in order to free up key staff,
- the contractor wishes to avoid liability for liquidated damages, and
- the client must have the building completed on time.

If the client wishes to investigate the possibility of achieving practical completion before the completion date, the architect / CA has the power under some standard forms of contract to invite proposals from the contractor as to how his can be achieved. When using JCT (16), upon receiving this request the contractor must either:

- provide the client with an acceleration quotation for accelerating the works identifying the time that can be saved, or
- explain why acceleration would be impractical.

If the contractor decides to provide an acceleration quotation, the following items must be clearly identified:

- the amount of time that could be saved,
- the adjustment to the contract sum that would be required split down into:
    - direct costs,
    - consequential loss and expense, and
    - the cost of preparing the quotation.

There isn't any published guidance on how to prepare an acceleration quotation and therefore it is up to the contractor to decide:

- what can be achieved,
- how it can be achieved, and
- how much it will cost.

The acceleration quote must be provided with twenty-one days and remain open for consideration for seven days, however in practice twenty-eight days may be too long and therefore these parameters may be varied by mutual consent.

It should be noted by project managers that accelerating the contract programme is not without its risks, which include the following considerations:

- It has been tested in the courts and found that contractors do not have a general duty to accelerate the works.
- What happens if the acceleration quotation is approved but acceleration is not achieved?
- Contractors could insert a caveat disclaiming responsibility for further delays.
- If the client requires guaranteed revised completion dates, the contractor may charge a premium.
- If the contractor is to accept all risks then the quotation should be in the form of a lump sum.
- If the client is to share / accept risks, then the quotation should be broken down.

How can acceleration be achieved?

In order to accelerate the works the employer may be able to change:

- the specification,
- the design, and
- the scope of the work.

However, it should be noted that the consequences of this could be:

- the client ends up with a project below expectations,
- there could be additional design fees,
- omitting significant sections of a project is rarely practical,

- deferring works until after handover is not strictly acceleration, and
- if work is omitted the contractor may claim for loss of profit.

The contractor may propose that the working hours are extended, but it should be borne in mind that this approach may not be very productive for a number of reasons:

- the extended hours would probably be undertaken as overtime, therefore this will be more expensive than basic plain time,
- the existing staff will be tired and possibly less productive,
- weekend working is even more expensive, and
- restrictions on noise levels / access may apply.

Another approach is to increase resources, that is to say increase the level of supervision, labour and plant. However it is important to keep the correct balance and avoid a disproportionate increase in labour without added plant for example. Another point to consider is whether there is the room on-site for increased resources, as increased resources may involve more storage, welfare and accommodation.

NEC4

When using NEC4, the process is generally similar to JCT except that:

- Unlike JCT (16) it is not up to the contractor to work out what acceleration can be achieved, rather it is the project manager who informs the contractor of the revised completion date.
- The contractor may provide a quote or decline to do so, giving reasons.

Insolvency / bankruptcy / sequestration

Construction is a risky business, as borne out by the Insolvency Service's official statistics, which reveals that there were 5603 insolvencies in 2018 in England and Wales in construction and supporting services. Similar statistics are available for sequestration in Scotland. Construction and property firms made up approximately 20 per cent of the total number of all businesses in England and Wales forced into liquidation and consequently project managers should be to alert to the possibility of contractors and / or subcontractors ceasing trading and the impact that this may have on the delivery of the project, particularly when market conditions are difficult with a shortage of work.

Insolvency is defined by the Insolvency Act 1986 and 2000 and the Insolvency Rules 2016. A company or individual is deemed to be insolvent when:

- they have insufficient assets to cover their debts or are unable to pay bills when they are due,
- a creditor to whom the company is indebted in a sum exceeding £750 then due has served on the company, by leaving it at the company's registered office, a written demand (in the prescribed form) requiring the company to pay the sum so due, and the company has for three weeks thereafter neglected to pay the sum or to secure or compound for it to the reasonable satisfaction of the creditor.

Sequestration is the seizure of property for creditors or the state. In the case of insolvency / bankruptcy / sequestration of the main contractor the project manager should ensure that the following tasks are carried out immediately:

- Withhold payments and the release of retention monies in accordance with the Housing Grants, Construction and Regeneration Act 1996 and the Local Democracy, Economic Development and Construction Act 2009.
- Lock the site and make it safe, and secure both contractors' and subcontractors' materials and plant. Note that position regarding ownership of materials off-site is different in Scotland.
- Make use of existing temporary site set up.
- Employ a security firm to safeguard the site and materials.
- Decide whether to retain scaffolding / structural issues, as this will probably be owned by a third party.
- Obtain legal advice or advice on termination of the contract.
- The contractors' insurance obligations could cease when the contract is terminated and therefore the level of insurance cover should be checked immediately and the appropriate action taken.
- Carry out an inspection of the site and record the progress accurately.
- Prepare a schedule of the works to complete as at the date of termination.
- Prepare a schedule of materials on-site and inspect materials off-site. If they are stored at the insolvent parties' premises and have been paid

for by the employer, make an arrangement to move them to alternative secure premises.

- Serve an urgent work notice and / or repairs notice as necessary to further safeguard or secure the site and environment.
- Under the Construction Design and Management (CDM) Regulations 2015, the client is required to ensure continuity and takes on the responsibilities of the main contractor for site safety from termination of the building contract until the appointment of a new main contractor.
- Advise on the options for completion of the project.
- Advise on the options for claims, such as bonds, guarantees, collateral warranties, loss and / or expense, prolongation, disruption, overvaluing of the works, professional indemnity claims for consultants, extensions of time, and liquidated and ascertained damages.
- Obtain and collate design information, specifications and drawings. Obtain any technical advice or reports on installed work and / or contractors' specifications.
- Advise a schedule of defects is made at the date of termination. This may involve some intrusive investigation if it is reasonably anticipated that works were not installed satisfactorily.
- Send enquiries and obtain quotations for the making good of defects on a competitive basis.
- Liaise with the insolvency practitioner.
- Advise as to whether professional indemnity insurance is being maintained or is at risk of lapsing.
- Write and issue a notional final account.
- Write and issue a completion final account.
- Complete general administration of the project, including financial monitoring and retendering.
- There will almost certainly be outstanding works that require completion and a tendering process should be put in place to select a new contractor to complete the works.

Suspension of payment

Payments that are due must not be delayed and also must not be unfairly reduced in an attempt to prepare for possible insolvency (i.e. not undervalued). Before insolvency is proven, delaying or reducing payments arbitrarily is unreasonable, is not authorised under the contract and can have a detrimental effect to the party being considered and also to the project as a whole. If the interim payments are undervalued, the contractor is denied their operating capital and may struggle to pay their debts. Conversely, overvaluation may in the case of subsequent insolvency involve the recovery of the overvalued monies which, under the circumstances, will be difficult to achieve. It may also lead to professional indemnity issues which can be avoided if work is properly and reasonably valued in the first place.

The following are telltale signs for a project manager that a contractor or subcontractor is heading into financially troubled waters:

- subcontractors and suppliers are not being paid, and
- material supply is interrupted as accounts with builder's merchants have been suspended.

Stage payments / cash flow projections

Most standard forms of contract have a provision to pay the contractor on a stage / interim payment basis. Without this facility the contractor would have to fund the project from their own sources which would have financial and logistical implications.

The client will need to know the amount and the timing of these payments in order to ensure that adequate funds are available, and a schedule of payments and cash flow projection should be prepared by the project manager.

The Housing Grants, Construction and Regeneration Act 1996, revised in 2011, is applicable to all contracts entered into after 1 October 2011. The client must provide both for stage payments and adequate mechanisms for determining payments. The Act requires that construction contracts must include the following provisions:

- payment by instalments – usually monthly,
- mechanisms to determine what amounts are due and when,
- prior notice of the amounts due and how they were drawn up,
- prior notification (seven days) of the intention to withhold payments – sometimes referred to as 'set off', giving the reasons and the amount. (this provision was introduced to stop the practice by some clients of withholding sums of money from sums certified by the CA without notice),
- suspension of work by the contractor for non-payment money due,
- all 'pay when paid' clauses are not allowed; this was a practice used by some clients of withholding money from contractors and subcontractors,

sometimes without due cause, on the basis that they will only be paid once the client or main contractor had been paid.

In addition to helping the client's financial planning, cash flow projections can also help the project manager in cost control of the project. As well as a cash flow projection for construction costs the client may also request a similar projection for professional fees. Figure 3.4 illustrates a typical lazy standard curve of a cash flow projection; starting slowly, peaking during the last third of the project before tailing off.

Supply chain relationships and management

A construction project team is usually a temporary organisation designed and assembled for the purpose of the particular project. It is made up of different companies and practices, which have not necessarily worked together before and which are tied to the project by means of varying contractual arrangements. This is what has been termed a temporary multi-organisation; its temporary nature extends to the workforce, which may be employed for a particular project, rather than permanently. These traditional design team / supply chain models are the result of managerial policy aimed at sequential execution and letting out the various parts of the work at apparently lowest costs.

Most of what is encompassed by the term supply chain management was formerly referred to by other terms such as 'operations management', but the coining of a new term is more than just new management speak, it reflects the significant changes that have taken place across this sphere of activity. These changes result from changes in the business environment. Most manufacturing companies are only too aware of such changes: increasing globalisation, savage price competition, increased customer demand for enhanced quality and reliability, etc. Supply chain management was introduced in order that manufacturing companies could increase their competitiveness in an increasingly global environment as well as their market share and profits by:

- minimising the costs of production on a continuing basis,
- introducing new technologies,
- improving quality, and
- concentrating on what they do best.

The contrast between traditional approaches and supply chain management can be summed up as shown in Figure 3.5.

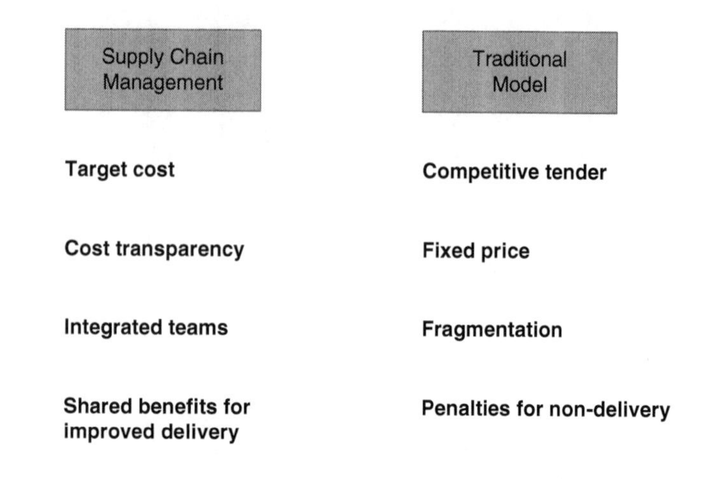

Figure 3.5 Supply chain management

Unlike other market sectors, because the majority of organisations working in construction are small, the industry has no single organisation to champion change. When a series of government reports called for a 30–50 per cent reduction in costs, the knee-jerk response from some quarters of the profession and industry was that cost = prices and it was impossible to reduce the prices entered in the bill of quantities by this amount, therefore the target was unrealistic and unachievable. But, reducing costs goes far beyond cutting the prices entered in the bill of quantities, it extends to the reorganisation of the whole construction supply chain in order to eliminate waste and add value. The immediate implications of supply chain management are:

- key suppliers are chosen on criteria, rather than job by job on competitive quotes,
- key suppliers are appointed on a long-term basis and proactively managed, and
- all suppliers are expected to make sufficient profits to reinvest.

How many project managers have asked themselves this question at the outset of a new project: '*What does value mean for my client?*' In other words, in the case of a new plant to manufacture, say, pharmaceutical products, what is the form of the built asset that will deliver value for money, over the life cycle of the building for that particular client? For many years, whenever clients have voiced their concerns about the deficiencies in the finished product, all too often the patronising response from the profession has been to accuse the complainants of a lack of understanding in either design or the construction process or both.

The answer to the value question posed above will of course vary between clients, a large multi-national manufacturing organisation will have a different view of value to a wealthy individual commissioning a new house, but it helps to illustrate the revolution in thinking and attitudes that must take place. In general, the definition of value for a client is '*design to meet a functional requirement for a through-life cost*'. Project teams are increasingly developing better client focus, because only by knowing the ways in which a particular client perceives or even measures value, whether in a new factory or a new house, can the construction process ever hope to provide a product or service that matches these perceptions. Once these value criteria are acknowledged and understood there are a number of techniques at their disposal in order to deliver to their clients a high degree of the feel-good factor. For example:

- measure productivity – for benchmarking purposes,
- measure value – demonstrating added value,
- measure outturn performance – not the starting point,
- measure supply chain development – check suppliers are improving as expected, and
- measure ultimate customer satisfaction – customers at a supermarket, passengers at an airport terminal, etc.

Of course, measuring value is extremely difficult to do.

What is a supply chain?

Before establishing a supply chain or supply chain network, it is crucial to understand fully the concepts behind and the possible components of a complete and integrated supply chain. The term supply chain has become used to describe the sequence of processes and activities involved in the complete manufacturing and distribution cycle – this could include everything from product design, through materials and component ordering, through manufacturing and assembly, until the finished product is the hands of the final owner. Of course, the nature of the supply chain varies from industry to industry. Members of the supply chain can be referred to as upstream and

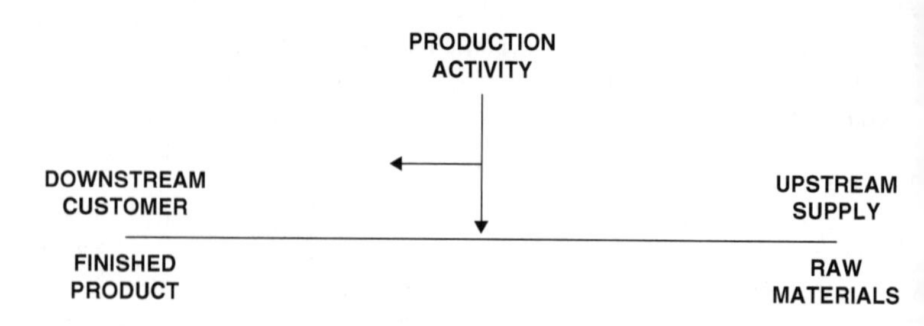

Figure 3.6 A supply chain

downstream supply chain members, as illustrated in Figure 3.6. Supply chain management, which has been practised widely for many years in the manufacturing sector, therefore refers to how any particular manufacturer involved in a supply chain manages its relationship both up- and downstream with suppliers to deliver cheaper, faster and better. In addition, good management means creating a safe commercial environment, in order that suppliers can share pricing and cost data with other supply team members.

The more efficient or lean the supply chain, the more value is added to the finished product. As if to emphasise the value point, some managers substitute the word value for supply to create the value chain. In a construction context supply chain management involves looking beyond the building itself and into the process, components and materials which make up the building. Supply chain management can bring benefits to all involved when applied to the total process, which starts with a detailed definition of the client's business needs which can be provided through the use of value management and ends with the delivery of a building providing the environment in which those business needs can be carried out with maximum efficiency and minimum maintenance and operating costs. In the traditional methods of procurement the supply chain does not understand the underlying costs, hence suppliers are selected by cost and then squeezed to reduce price and whittle away profit margins. The clear disadvantages are:

- bids based on designs to which suppliers have no input, hence buildability is compromised,
- low bids always win,
- since this is unsustainable, costs need to be recovered by other means,

- margins are low, so no money to invest in development, and
- suppliers are distant from the final customer so take limited interest in quality.

The traditional construction project supply chain can be described as a series of sequential operations by groups of people or organisations.

Supply chains are unique, but it is possible to classify them generally by their stability or uncertainty on both the supply side and the demand side. On the supply side, low uncertainty refers to stable processes, while high uncertainty refers to processes which are rapidly changing or highly volatile. On the demand side, low uncertainty would relate to functional products in a mature phase of the production life cycle, while high uncertainty relates to innovative products. Once the chain has been catagorised, the most appropriate tools for improvement can be selected.

The construction supply chain is the network of organisations involved in the different processes and activities that produce the materials, components and services that come together to design, procure and deliver a building. Traditionally it is characterised by lack of management, little understanding between tiers of other functions or processes, lack of communication and a series of sequential operations by groups of people who have no concern about the other groups or client. Figure 3.2 illustrates part of a typical construction supply chain, although in reality many more subcontractors could be involved. The problems for process control and improvement that the traditional supply chain approach produces are related to:

- the various organisations come together on a specific project, at the end of which they disband to form new supply chains,
- communicating data, knowledge and design solutions across the organisations that make up the supply chain,
- stimulating and accumulating improvement in processes that cross the organisational borders,
- achieving goals and objectives across the supply chain, and
- stimulating and accumulating improvement inside an organisation that only exists for the duration of a project.

However, supply chain management takes a different approach:

- Prices are developed and agreed, subject to an agreed maximum price, with overheads and profit ring-fenced. All parties collaborate to drive

down cost and enhance value, for example through the use of an incentive scheme.
- With costs determined and profit ring-fenced, waste can now be attacked to bring down price and add value with an emphasis on continuous improvement.
- As suppliers account for 70–80 per cent of building costs they should be selected on their capability to deliver excellent work at competitive rates.

Suppliers should be able to contribute new ideas, products and processes, build alliances outside of project and be managed so that waste and inefficiency can be continuously identified and driven out.

The philosophy of integrated supply chain management is based upon defining and delivering client value through established supplier links that are constantly reviewing their operation in order to improve efficiency. There are now growing pressures to introduce these production philosophies into construction and it is quantity surveyors, with their traditional skills of cost advice and project management, who can be at the forefront of this new approach. For example, the philosophy of Lean Thinking, which is based on the concept of the elimination of waste from the production cycle, is of particular interest in the drive to deliver better value. In order to utilise the Lean Thinking philosophy, the first hurdle that must be jumped is the idea that construction is a manufacturing industry which can only operate efficiently by means of a managed and integrated supply chain. At present the majority of clients are required to procure the design of a new building separately from the construction, however as the subsequent delivery often involves a process where sometimes as much as 90 per cent of the total cost of the completed building is delivered by the supply chain members, there would appear to be close comparisons with, say, the production of a car or an aeroplane.

The basics of supply chain management can be said to:

- determine which are the strategic suppliers, and concentrate on these key players as the partners who will maximise added value,
- work with these key players to improve their contribution to added value, and
- designate these key suppliers as the 'first tier' on the supply chain and delegate to them the responsibility for the management of their own suppliers, the 'second tier' and beyond.

To give this a construction context, the responsibility for the design and execution of, say, mechanical installations could be given to a 'first tier' engineering specialist. This specialist would in turn work with its 'second tier' suppliers as well as with the design team to produce the finished installation. Timing is crucial, as first tier partners must be able to proceed confident that all other matters regarding the interface of the mechanical and engineering installation with the rest of the project have been resolved and that this element can proceed independently. That said, at least one food retail organisation using supply chain management for the construction of its stores still places the emphasis on the tier partners to keep themselves up to date with progress on the other tiers, as any other approach would be incompatible with the rapid timescales that are demanded.

Despite the fact that on the face of it, certain aspects of the construction process appears to be a prime candidate for this approach, the biggest obstacles to be overcome by the construction industry in adopting manufacturing industry-style supply chain management are:

- Unlike manufacturing, the planning, design and procurement of a building is at present separated from its construction or production.
- The insistence that, unlike an aeroplane or cars, every building is bespoke, a prototype, and therefore is unsuited to this type of model, or for that matter any other generic production sector management technique. This factor manifests itself by:
  - geographical separation of sites that causes breaks in the flow of production,
  - discontinuous demand, and
  - working in the open air, exposed to the elements – can there be any other manufacturing process, apart from shipbuilding, that does this?
- Reluctance by the design team to accept early input from suppliers and subcontractors, and unease with the blurring of traditional roles and responsibilities.

There is little doubt that the first and third hurdles are the result of the historical baggage outlined in Chapter 1 and that, given time, they can be overcome, whereas the second hurdle does seem to have some validity despite statements from the proponents of production techniques that buildings are not unique and that commonality even between apparently differing building types is as high as 70 per cent (Ministry of Defence, 1999). One of the main elements of supply chain management, Just in Time (JIT), was reported to have started in the Japanese shipbuilding industry in the mid-1960s, the very industry that opponents of JIT in construction quote as an example of where, like construction, supply chain management techniques are inappropriate. Therefore, the point at which any discussion of the suitability of the application of supply chain management techniques to building has to start with the acceptance that construction is a manufacturing process, which can only operate efficiently by means of a managed and integrated supply chain. One fact is undeniable – at present the majority of clients are required to procure the design of a new building separately from the construction. Until comparatively recently, international competition, which in manufacturing is a major influencing factor, was relatively sparse in domestic construction in major industrialised countries.

INSURANCES – INJURY, DAMAGES AND INSURANCE

Section 6 of the JCT (16) deals with indemnities and insurances to persons, property, the works and, in addition, professional indemnity insurance in cases where contractor design in involved.

Injury to persons and property

The clauses referring to injury to persons and property have been significantly reworded in JCT (16). Contractors should confirm that apprenticed staff are covered and consider whether self-employed individuals are covered. Clauses 6.1 and 6.2 require the contractor to indemnify the employer for any injury to persons or property that occurs during the carrying out of the works. This clause has the effect of protecting the employer from any claim that may be made for any injury to persons or property. It should be noted that some items are specifically excluded from the indemnity provisions and these are set out in clause 6.3, namely loss or damage to existing structures or their contents, for example sections of the works for which a practical completion certificate has been issued, which becomes the property of the employer.

Section 6 of JCT (16) continues with a requirement for the contractor to take out insurance to cover the items covered in clauses 6.1 and 6.2 which must be produced within seven days of the request. The contractor must allow the employer to inspect the insurance policy and if it is considered to be inadequate, then the employer can take out his / her own policy and deduct the cost from any money due to the client.

Insurance of the works

Insurance of the works is dealt with by Section 6 of JCT (16) and gives the parties to the contract three options. Only one option should be used. Options A and B are for new buildings, whereas Option C is for alterations and works to existing buildings.

Options A and B are similar, except that in Option A it is the contractor who takes out the policy for all risks insurance, whereas in Option B it is the employer who takes out the policy, again in joint names. Therefore the main difference between the two options is that in the case of a claim it is either the contractor or employer who receives the insurance monies and arranges to make good any damage, etc.

It is a sign of the times that a major change in JCT (16) is a definition of terrorism which greatly widens the range of events that ought to be covered by insurance.

NEC4 Section 8 sets out the insurance liabilities for the contractor and the client. There is no distinction between new and existing buildings.

Contractor-designed portion professional indemnity insurance

Clauses 6.15/16 of JCT (16) are a comparatively new clauses to the JCT contract and require the contractor to maintain professional indemnity insurance in respect of any contractor's designed portion and provide proof of a policy on request.

Joint Code of Practice: Fire prevention on construction sites (Ninth Edition)

Clauses 6.17–6.20 of JCT (16) refer to the Joint Fire Code. The code applies to activities carried out prior to and during the procurement, construction and design process, not the completed structure, and should be read in conjunction with all current legislation.

The object of this code is the prevention of fires on construction sites. It is claimed that the majority of fires can be prevented by designing out risks, taking simple precautions and by adopting safe working practices. All parties involved must work together to ensure that adequate detection and prevention measures are incorporated during design and contract planning stages, and that the work on-site is undertaken to the highest standard of fire safety, thereby affording the maximum level of protection to the building and its occupants.

The code is voluntary, but if applied the contractor and employer are bound to comply with it.

Bonds

A bond may be thought of as a guarantee of performance. The JCT (16) includes the provision to execute three forms of bond:

- **Advanced payment bond** – advanced payments are not as common in the UK as, say, France, where they are commonplace. In the event that the contractor requires an advanced payment from the employer, prior to work commencing on-site, an advanced payment bond must guarantee to repay the advance in the case of default by the contractor, JCT (16) clause 4.7 and Schedule 6.1. NEC4 Option X14 allows for the provision of advanced payments to a contractor with or without a bond.
- **Bond for off-site materials and / or goods** – there may be occasions where, for example, it may be necessary to purchase materials and goods in advance and to store them off-site. The goods and materials in question may be expensive or delicate or both, and therefore it is inappropriate to store them on-site. Nevertheless, the contractor applies for payment and, in these circumstances, a bond to cover the cost of the materials or goods should they be damaged or lost. JCT (16) Schedule 6.2 applies.
- **Retention bond** – retention, normally at 3 per cent, is deducted from all interim payments made to the contractor and held by the employer until practical completion and final account stages. If the contractor provides a retention bond then the retention deduction may be waived. JCT (16) clause 4.18 applies and Schedule 6.3.

In addition, other common forms of bond are:

- **Performance bond** – a performance bond is required to guarantee the performance of the contract during the works. In value terms it is usually equal to up to 10 per cent of the value of the contract. The purpose of the performance bond is to reimburse the client in the event that contractor does not proceed diligently. NEC Option X13 allows for the use of a performance bond.
- **Tender bond** – a bond may be required by a client to ensure that contractors who express an interest in submitting a bid for a project are bona fide. The

bond fund may be used in the event that a contractor either fails to submit a bid or fails to enter into a contract after being selected.

Guarantees – collateral warranties

Generally, a warranty is a term of a contract, the breach of which may give rise to a claim of damages but not the right to treat the contract as repudiated. It is therefore a less important term of the contract, or one which is collateral to the main purpose of the contact, the breach of which by one party does not entitle the other to treat his obligations as discharged.

Undertakings may be given that are collateral to another contract, that is running side by side. They may be independent of the other contract because they cannot be fairly incorporated, or the rules of evidence hinder their incorporation, or because the main contract is defective in some way. A warranty is a term of a contract, the breach of which may give rise to a claim of damages. A transaction between two parties may be of particular concern or affect to the performance of a third party. A collateral contract may be entered into between the third party and one of the original parties. This may be a useful device for avoiding privity of contract.

Increasingly there are a number of parties with financial stakes in the success of a construction project (e.g. funders, tenants, and purchasers) who are not party to the building and other associated contracts. In the event of a third party suffering loss arising from the construction project, in the absence of a direct contract, the only remedy is a claim in the tort of negligence. However, since the 1980s the courts have severely restricted the scope of negligence claims. In order to plug this contractual gap the parties to the contract may decide to make use of collateral warranties. These are contracts in which the person or firm doing the work (the warrantor) warrants that they will properly carry out their obligations under the main contract. Therefore, in the case of defective workmanship or the like, the warrantor may be sued. As with standard forms of contract there are many standard forms of collateral warranty, included those published by the JCT. It is also possible to use bespoke forms of collateral warranty for those who fear that their right to pursue a warrantor is limited by clauses in the standard forms. A classic model for the use of collateral warranties is when management contracting procurement strategy is used and the package contractors have responsibility for elements of the design. Under these circumstances, the client would procure a series of collateral warranties with the relevant package contractors. The principal disadvantage of using collateral warranties, as in the case of management contracting, is the amount of time and bureaucracy involved in procuring warranties from twenty to thirty separate organisations. It should be borne in mind that a collateral warranty should not expose the warrantor to any greater exposure than they had under the original contract and the extent of liability, usually restricted to the cost of remedial works only.

Collateral contracts

In theory, the Contracts (Rights of Third Parties) Act 1999, that came into force in May 2000 provides an alternative to collateral warranties. One of the intentions of the Act was to reduce the need for collateral warranties but initial take up of the Act has been disappointing. However, when the JCT considered the provisions of the 2005 suite of contracts it was decided that provisions should be incorporated for a contractual link between the main contractor, funders, etc. The majority of new 16 forms now include an obligation to provide warranties and a mechanism for invoking the Act. The new forms also allow for the calling of subcontract warranties in favour of such third parties and the employer. Both the JCT (16) (Section 7) and the JCT Design and Build (16) contain provision for collateral warranties to third parties, subcontractor collateral warranties and third-party rights.

Where there is no collateral contract found, a plaintiff may still sue in negligence. However, a claim may be purely economic and this may well prove fatal in establishing a duty of care. Further, claims in contract by implied terms (for example implied by the Sale of Goods Act 1979) are normally strict, that is have no defence, but a claim in negligence will require proof of fault.

Collateral warranties between employer and subcontractor

When a subcontractor enters into a domestic contract with a main contractor there is no contractual relationship between the employer and the subcontractor. The employer could only sue the subcontractor in tort and would have to prove that a duty of care was owed. A collateral warranty between employers and subcontractors allows the employer to sue the subcontractor for any breach of the warranty's conditions, which commonly include promises on the part of the subcontractor to achieve a standard of design and workmanship as specified by the employer. The terms of the warranty may impose whatever liabilities and responsibilities the employer considers appropriate, so long as the subcontractor, being aware of such terms, is willing to tender and enter into a subcontract for the relevant work.

It is important, but nevertheless sometimes overlooked, to ensure that both the obligation to enter into the warranty and its full wording form part of the legal obligations set down by the terms of the contract between the main contractor and the subcontractor or professional party. Warranties will also address the matter of deleterious materials, to ensure that such materials are not specified or employed in the works.

Most collateral warranties include provisions for the benefit of the warranty to be assigned by the employer to a third party, such as a purchaser or tenant. Indeed such third parties taking a legal interest in a building require such a warranty, so as to provide themselves with redress against a contractor or designer as a result of defects appearing within a period of time, commonly after twelve years from the completion of the original works.

On completion, either partial, sectional or for the total project, the client is able to occupy and use the building and the post-construction phase begins.

4

Post-construction / OGC Gateway 4–5 / RIBA Plan of Work Stage 6

What is the difference between post-construction and occupancy phases from the project management perspective?

- At the post-construction stage the construction work is approaching completion and the project manager is concerned with organising the client taking over possession and moving into the new facility as smoothly as possible.
- Occupancy involves the optimisation and project management of the building once the owner / client has moved in and starts to operate / run their business (see Chapter 5).

POST-CONSTRUCTION

Towards the end of the construction phase (RIBA Plan of Work Stage 5) there will be a point when the client can take possession of the facility; this point is generally referred to as practical completion or handover. The word practical is used in JCT (16) as there is a realisation that when dealing with a complex process, such as constructing a new building or refurbishing an existing building, it is not possible to completely finish every task at a prescribed point and that there will be outstanding items left to be done. Nevertheless, these outstanding items do not stop the client taking possession of the building and using it for its intended purpose. At the point of practical completion the rectification period commences when the outstanding items are completed and defective work rectified by the contractor. Half of the retention fund will be released to the contractor at this stage. The project manager should note however that there is no legal basis for the application of this convention. Even if the works can be 'beneficially occupied', unless the contract provides otherwise the client is not obliged to take possession of the project. A similar approach is taken by NEC4 Section 3, in that completion can be achieved even if minor defects exist.

The project manager should agree a handover strategy with the client which should include:

- a schedule of responsibilities,
- the transfer of project documentation including operating and maintenance manuals, and
- processes for commissioning and rectifying latent defects, that is to say defects that are not apparent at the time of occupation.

If BIM has been used for the project, much of this information, such as project and facilities management documentation, could be embedded in the BIM model.

BIM offers potential efficiency gains to the operational phase of a building by helping FM managers to:

- understand what components have been used to construct the building and where they are located,
- understand and manage energy use more effectively / efficiently,
- appreciate life cycle costs by giving a more complete picture,
- understand how to adapt systems when reconfiguration of a building is required, and
- greatly simplify maintenance.

What happens at practical completion?

The project manager should be aware and keep the client informed that at practical completion the following takes place:

- The contract administrator issues a Certificate of Practical Completion stating the date at which certain responsibilities are transferred to the client. The certificate should include a list, often referred to as a snagging list, containing items still to be completed or are outstanding.
- The responsibility for insurance of the works and security is transferred to the client.
- Half of the retention fund is released to the contractor, typically 1.5 per cent of the certified sums.
- The rectification period commences, during which defects and outstanding items are to be completed.

- For smaller projects, a rectification period of three to six months may be appropriate, while for larger projects a period of six to twelve months (minimum) is recommended, although eighteen- to twenty-four-month periods are not uncommon. The project manager should check with the architect / CA to determine the most suitable period of time, and to arrange a process for the recording and referral of issues for action by the contractor.
- In the case of sectional completion, the release of retention monies should be calculated in proportion to the value of section of the works completed.
- There is a requirement for the final account to be prepared.
- The length of the rectification period, as well as any details of phased handover, will be entered into the conditions of contract.

The rectification period marks the start of the period when the final account is prepared, although much of the spade work should have already been carried out by the client's quantity surveyor and the contractor. The project manager should arrange to receive regular updates on the progress of the final account. The final account is a reconciliation of the tender price, to include the cost of variations issued during the works, together with any agreed claims. The format and detail required for the final account will vary according to the client, from the public sector where a high degree of accountability is usually required, to the private sector where so much detail will not normally be usual.

During the construction stage the project manager should arrange to receive regular, monthly or quarterly financial statements from the quantity surveyor, therefore at the final account stage the amount of the final account should be known to the project manager and the client. See Chapter 3 and Appendix A.

It is not uncommon for there to be provision in the contract for sectional or partial practical completion as, in addition to the whole of the works being handed over to the client at practical completion, the client may ask for partial or sectional completion when a self-contained portion of the work is handed over to the employer, while work continues on the remaining sections of the project. This may be particularly useful in very large or mixed-use developments. In such cases there may be several periods of final measurement, releases of retention, etc. and they require careful monitoring by the project manager, particularly where subcontractors are involved. It is important that the project manager understands the definition of completion in terms of a building project.

What does completion mean?

With general commercial contracts, completion is when all the obligations have been satisfied; however, applying commercial criteria to construction projects could expose the contractor to liabilities for damages. Under JCT (16), completion is generally referred to as practical completion which has been interpreted as allowing a state of less than full completion. Under NEC4, formal completion is defined as when all the work required by the works information is completed and that all notified defects that would prevent the client from using the project are corrected. This implies that completion can be achieved even if there are defects within the works, providing that the defects do not prevent the client from using the works. Having said this, there is no legal basis for determining practical completion, and if the works are not finished then the client is not obliged to take possession. In the case of construction contracts, it is thought that imposing onerous criteria on construction projects could be seen as unfair, especially when minor but time-consuming tasks are outstanding.

Sectional completion

Often a client will want to take possession of sections of the work without the whole of the works being completed, and if this is the case then this will usually be anticipated and stated in the contract documents. For example, in the case of a mixed-use development, the retail units could be available prior to the other parts of the development.

Partial completion

Sometimes a client may want to take possession of part of the works without previous arrangements in the contract documents. Under these circumstances partial completion under JCT (16) cannot take place without the consent of the contractor, but may not be unreasonably withheld. Partial possession usually anticipates discrete parts of the works being taken into possession of the client at an early stage.

Latent and patent defects

Latent defects are defects that are not apparent at completion but subsequently become apparent, for example failure in structural concrete due to inadequate reinforcement, whereas patent defects are defects that are known at the completion. It is possible to for the client to insure against the impact of latent defects for a period of several years after completion; several insurance companies will write policies to cover buildings against the impact of latent defects for a period of ten years after completion. The big advantage of latent defects insurance is that any defects can be immediately rectified instead of waiting to establish liability before remedial works can be commenced.

Practical completion, therefore, marks the end of the main construction operations and the project manager needs to ensure that a number of procedures are completed as follows:

- Ensure that the architect / CA has inspected the works and issued a Certificate of Practical Completion. This is important as the date on the certificate determines the commencement of the rectification period and marks the start of the preparation of the final account. An agreed snagging list of items still to be completed by the contractor should be attached to the certificate of practical completion (see Appendix B).
- In the case sectional of partial or completion, the project manager should ensure that the correct parts of the project are clearly identified.
- Insurance should be put in place by the client to cover the risks which, up to that point, have been covered by the contractor.
- Upon the issue of the Certificate of Practical Completion, the project manager should ensure that the final account is prepared by the client's and contractor's professional advisors. Although some forms of contracts stipulate a time frame for this process – nine months in the case of JCT (16) – there is no penalty for non-completion and therefore it is particularly important for the project manager to monitor the process.
- Although a list of defects is attached to the Certificate of Practical Completion, new defects will come to light during the rectification period itself and a system should be put in place to record these defects and add them to the snagging list for the attention of the contractor.
- When all defects are rectified, the project manager should ensure that the architect / CA completes a final inspection of the works and issues a final certificate. Upon the issue of the final certificate all works are deemed to be to the satisfaction of the architect / CA.

At practical completion a number of key documents are handed to the client / sponsor, for example:

- drawings,
- operating and maintenance manuals,
- warrantees,
- commissioning documents, and
- health and safety documents.

However, this process will vary if:

- BIM has been used for the project, and / or
- Soft Landings are being used (see Chapter 2 and below).

If BIM has been used for the project, then instead of hard copies of drawings, etc. physically being handed over to the client, documents can be loaded onto the model. Therefore, when the client wishes to refer to maintenance manuals, for example, it is only necessary to refer to the appropriate section of the model.

TAKING POSSESSION

Client preparations for occupying the new facility should enable them to move the project from the regime of a building site to an occupied working facility whilst incurring minimal disruption and cost to the business. The client should be advised to appoint a senior member of staff responsible for moving who can stand their ground under pressure from various elements of management who may fight their corner during the stress and upheaval of moving. The job is one that requires skillful co-ordination of a multitude of time-consuming tasks. On larger projects they are likely to require a team under them including an accommodation manager and perhaps a facilities manager. The person responsible for moving should prepare an operational policies document setting out the detailed plan for how the building will be occupied and used. They should also prepare a migration strategy (see migration strategy later in this chapter).

The operational policies document may be based on information from:

- the business case,
- the project brief,
- the developed design,
- corporate planning strategies, and
- the building user's guide.

The operational policies document might include:

- room data sheets,
- space planning information,
- furniture, desk and information and communications technology (ICT) allocations,
- equipment schedules,
- schedules of items that will be leased or purchased,
- requirements for consumables such as stationery and sanitary supplies,
- transport and parking policies for staff, visitors, VIPs, goods in and waste out, public and private buses and drop off facilities, and
- operational services requirements (see below).

A strategic decision should be made in the development of operational policies as to which services might be outsourced and which will remain under in-house direct control. Items that could be outsourced (if applicable) may include:

- reception and telephony,
- security,
- cleaning,
- facilities management,
- information and communications technology (ICT) support,
- catering,
- waste management,
- landscaping and ground maintenance,
- transport and courier services, and
- maintenance and servicing of equipment.

It may be beneficial to the overall programme for the client to have use of certain areas of the building prior to practical completion (for example so that ICT suites can be equipped). Ideally this requirement for phased completion should be written into the building contract, otherwise it will need to be negotiated with the contractor. More often than not it will become apparent that since the original project brief was prepared there have organisational changes and technological advances that necessitate changes to the design and installation works. This, combined with furniture, ICT equipment, fixtures, fittings, art, shelving, vending machines and so on, will result in a schedule of works necessary prior to occupation that can cost as much as 3 per cent of the construction budget. It is not unusual to package this work into an occupational services contract separate from the main building contract to take place after practical completion but prior to occupation. This additional work needs to be defined, costed, tendered and the contract let with a reasonable period for mobilisation and pre-ordering so it can commence as soon as the building is handed over and is often carried out under the supervision of the facilities manager.

To note

- The client should have appointed an in-house or outsourced engineering team to witness testing and commissioning and to take over the running of the services as soon as practical completion is certified.
- The client also needs to ensure that funds are available to meet the release of 50 per cent of the retention fund upon practical completion.
- Utility and fuel supplies need to be tendered or negotiated prior to occupation.
- Training of staff and familiarisation with new systems and space usage prior to occupation is an essential part of pre-planning.
- The client should also check compliance with planning conditions that have to be satisfied prior to occupation, that the building control inspector has inspected and approved the works and that appropriate insurance is in place.

FACILITIES / DATA MIGRATION

Data migration is the process of transferring systems from an existing facility or building to a new facility or building. Organisations planning a data migration should consider which style of migration is most suitable for their needs. They can choose from several strategies, depending on the project requirements and available processing windows, but there are two principal types of migration:

- **Trickle migrations** take an incremental approach to migrating data. Rather than aiming to complete the whole event in a short time window, a trickle migration involves running the old and new systems in parallel and migrating the data in phases. This method inherently provides the zero downtime that mission-critical applications requiring 24/7 operation need. A trickle migration can be implemented with real-time processes to move data, and these processes can also be used to maintain the data by passing future changes to the target system. Adopting the trickle approach does add some complexity to the design, because it must be possible to track which data has been migrated. If this is part of a systems migration, it may also mean that source and target systems are operating in parallel, with users having to switch between them, depending on where the information they need is currently situated. Alternatively, the old systems can continue to be operational until the entire migration is completed, before users are switched to the new system. In such a case, any changes to data in the source systems must trigger remigration of the appropriate records so the target is updated correctly.

- **Big bang migrations** involve completing the entire migration in a small, defined processing window. In the case of a systems migration, this involves system downtime while the data is extracted from the source systems, processed, and loaded to the target, followed by the switching of processing over to the new environment. This approach can seem attractive, in that it completes the migration in the shortest possible time, but it carries several risks. Few organisations can live with a core system being unavailable for long, so there is intense pressure on the migration, and the data verification and sign-off are on the critical path. Businesses adopting this approach should plan at least one dry run of the migration before the live event and also plan a contingency date for the migration in case the first attempt has to be aborted. The old systems can continue to be operational until the entire migration is completed, before users are switched to the new system. In such a case, any changes to data in the source systems must trigger remigration of the appropriate records so the target is updated correctly.

There have been several widely reported cases of organisations, particularly high street banks, attempting big bang IT migration of customers' account details without sufficient thought to the 'what if' question – that is to say 'what if things don't go according to plan?', with disastrous results, leaving customers unable to access their accounts for days or even weeks.

A data migration project typically starts with a broad brief from the business to the IT team that leads to a technically-focused migration in which more data is moved than necessary, at a greater cost over a longer period of time than was forecast, resulting in multiple revisions at numerous stages.

In information technology, migration is the process of moving from the use of one operating environment to another operating environment that is, in most cases, is thought to be a better one. Migration can involve moving to new hardware, new software, or both. Migration can be small-scale, such as migrating a single system, or large-scale, involving many systems, new applications or a redesigned network.

One can migrate data from one kind of database to another kind of database. This usually requires converting the data into some common format that can be output from the old database and input into the new database. Since the new database may be organised differently, it may be necessary to write a program that can process the migrating files. Migration is also used to refer simply to the process of moving data from one storage device to another.

HANDOVER AND OPERATION

Handover schedule

Once the client is certain that the project will proceed, that is to say at an early point in the project programme, a senior person responsible for moving should be appointed.

After preparing a policy for occupation, setting out how the facility will be used, the director and their team should prepare a migration strategy setting out the procedures for moving in such a way as to minimise disruption whilst allowing the efficient reuse of assets from any existing facilities.

This migration strategy might include:

- a detailed, phased, logistical programme for purchasing or moving of furniture and equipment,
- a detailed programme for moving or recruiting staff,
- requirements for the hire of temporary equipment,
- removal contracts,
- setting up a help desk with a rapid response team,
- postal and information and communications technology (ICT) arrangements to ensure continuity of communication (including transfer of hardware),
- setting up 'goods in' and 'dispatch' rooms, a post room and an information and communications technology support centre,
- catastrophe planning for fire or flood,
- staff transportation strategy,
- parking allocation,

- access for consultants, contractors and suppliers for summer and winter checks of building services systems and environmental conditions (which can only be properly carried out in a fully operational building),
- room allocation,
- signage,
- catering facilities and environmental health approval of kitchen areas,
- liaison with emergency services,
- stocking and storage of goods and consumables,
- communications between facilities during the move,
- installation of existing equipment requiring electrical, drainage, extract or cooling services, such as vending machines or fume cupboards,
- a risk schedule with mitigation measures (for example the absence of key personnel, late building handover, alarm activation, interruption of power or water supply, etc.).

In addition, the client may also need to put procedures in place to move some of its staff and equipment so that it can continue to operate effectively during construction.

POST-PROJECT REVIEW

A post-project review may begin during the rectification period, that is, the period between practical completion, when the client may take possession of some or all of the project, and the making good of any defects. When the development is first occupied by the client, it is important for the project manager to visit the site immediately to identify any issues that need to be addressed quickly. It can be beneficial to establish a help desk and rapid response team to resolve issues as they arise if the size of the project warrants this.

A post-project review is undertaken to evaluate the effectiveness and efficiency of the project delivery process. To undertake a post project review, it is important for the project manager to seek the views of contractors, designers, suppliers and the client about how well the project was managed. This may include assessments of how well the delivery of the project performed against key performance indicators such as:

- quality of briefing documents,
- effectiveness of communications,
- performance of the project team,
- quality issues,

- health and safety issues,
- certification,
- variations,
- claims and disputes, and
- collaborative practices.

An evaluation can then be made of what lessons can be learned from the approach taken and an assessment and lessons learned report prepared.

END OF CONTRACT REPORT

Once the defects liability period has ended and the final account has been agreed, it may be advisable for the project manager to prepare an end of contract report. On a traditional contract, an end of contract report is a commentary or overview of the history of the main contract and can be useful for a number of reasons:

- In the public sector, public scrutiny of a project can lead to questions of audit, proprietary and transparency long after the project is completed.
- On all projects, unanticipated legal proceedings can require the history of a project to be dissected.
- It can provide a useful reference document setting out the contractor's performance, which can be helpful when considering whether to employ that contractor again.

An end of contract report may include the following as appropriate:

- **Contractor design obligations and performance:**
  - progress measured against contract programme,
  - adherence to design concept,
  - level of BIM competency,
  - co-ordination with others in relation to things such as setting out and interfaces,
  - statutory approvals and independent design checks,
  - design faults,
  - nature of variations, and
  - adequacy of resources.
- **Off-site fabrication / MMC:**
  - progress against programme,
  - manufacturing errors, omissions or faults,

    - variations and scope reductions or increases,
    - percentage of work against factory output,
    - resources employed,
    - suppliers and subcontracts, and
    - payments for off-site materials.
- **Site works:**
    - management resources,
    - progress against programme,
    - site co-ordination and efficiency,
    - labour or material shortages,
    - subcontractor performance,
    - progress photographs and installation records,
    - samples and testing,
    - condemned work,
    - nature of variations,
    - handover documentation, and
    - defects and snagging.
- **Contractual:**
    - reconciliation of final account against contract sum,
    - contingency (risk allowance) expenditure,
    - delays or disruption,
    - details of extensions of time,
    - details of claims and settlements,
    - details of liquidated and ascertained damages,
    - disputes proceedings,
    - insurance claims,
    - signed contract documents, and
    - meeting minutes.

FACILITIES MANAGEMENT

Use and aftercare – Soft Landings / Government Soft Landings

The BRISA Soft Landings framework's success prompted the government to develop its own interpretation to suit public sector priorities. Government Soft Landings broadly follows the same core principles. Key performance metrics of FM running costs and workplace efficiency are added to the operational energy and occupant satisfaction criteria in the original BRISA 2009 document.

The RIBA Plan of Work (2015), is already aligned with regulatory requirements of BIM and sustainability. But perhaps the most important aspect of the revision is the reinforcement of feedback within all its seven stages, and its specific reference to Soft Landings

The term Soft Landings refers to a strategy to ensure the transition from construction to occupation is as seamless as possible and that operational performance is optimised. This should include agreement to provide the information required for commissioning, training, facilities management and so on, and increasingly will include requirements for Building Information Modelling (BIM).

Although handover of the project to the client / sponsor is the critical point for the strategy known as Soft Landings, it is very important that the project manager ensures that Soft Landings are introduced and implemented from the very early stages of the design sequence, and it is for this reason that this topic is discussed here. To ensure that a Soft Landings strategy is implemented properly from the outset, it may be appropriate to appoint a Soft Landings champion to oversee the strategy. Facilities managers should also be involved from the early stages. Soft Landings documentation extends the duties of the team during handover and the first three years of occupation.

Soft Landings and Government Soft Landings (GSL) are a comparatively new approach and are focused on the point in a project when the client takes possession of the facility where traditionally the project manager, design team and contractor walk away, with the exception of contractual obligations relating to the rectification period. Soft Landings is a joint initiative between BSRIA (Building Services Research and Information Association) and UBT (Usable Buildings Trust) and was devised in the late 1990s. It is an open-source framework available to use and adapt free of charge from BSRIA. When put side by side, there is considerable similarity between GSL and BSRIA Soft Landings, the main divergence being that GSL are more prescriptive in their approach. In 2018 BSRIA Soft Landings was revised with a number of major changes including the adoption of six phases to replace the original five stages as follows:

1. Inception and briefing.
2. Design.
3. Construction.
4. Pre-handover.
5. Initial aftercare.
6. Extended aftercare and post occupancy evaluation.

The change was made in order to distance Soft Landings from the stages of existing plans of work.

Project managers should be aware that the Cabinet Office announced that GSL would be gradually introduced alongside BIM for the government estate from 2016 as part of a range of new regulatory measures. However, for some people, GSL has been an opportunity lost in the headlines of the mandate of BIM in 2016.

GSL, launched in 2011, goes much further than BRSIA, as it tackles three aspects of sustainability – environmental, economic and social (which includes functionality and effectiveness) – by setting and tracking targets. By using GSL, government departments will be required to define a series of high-level outcomes at the beginning of a project. GSL are tied into BIM data drops, which are in turn tied into the RIBA Plan of Work Stages and OGC Gateway Processes outlined below. Rather than issuing stage reports, data drops (or information drops) will take place. At present there is no clear definition regarding the specific information that is issued as part of a stage report. It is the government's intention that this changes and work is underway to clarify the information required at each data drop, with these being aligned to the project stages. To ascertain the right level of information, the government is considering the questions that need to be answered at each stage, which will enable the construction industry to consider what the building model must contain and its level of refinement at a given stage.

- Data Drop 1: Model – Requirements and Constraints.
- Data Drop 2: Model – Outline Solution.
- Data Drop 3: Model – Construction Information.
- Data Drop 4: Model – Operations and Maintenance Information.
- Data Drop 5 (and subsequent drops): Model represents Post Occupancy Validation Information and Ongoing Operation and Maintenance.

Stage 1 – Inception and briefing

More time for constructive dialogue between the designer, constructor, client and end user and FM provider / caretaker. A clear brief is essential. Roles and responsibilities among the client and construction team need to be spelled out to show up any gaps, and lessons from previous projects need to be shared openly. To ensure the design meets operational needs, the facilities management team should be involved at the early stages of design and the design team needs to agree how to measure performance in use.

Stage 2 – Design development and review

Brings the entire project team together to review insights from comparable projects and details how the building will work from the point of view of the manager, individual users and FM provider / caretaker. It is important to design for buildability, usability and manageability. Designers need to consider budgets and the technical expertise of the occupier. Peer reviews by independent experts can pinpoint problems, and design reviews should include people with different jobs and levels of seniority. The contract documentation needs to reflect the Soft Landings approach.

Stage 3 – Pre-handover

Enables end users to spend more time on understanding interfaces and systems before occupation. The main purpose of this stage is to make sure that the building is ready for operations. Problems that occur after handover can often be tracked back to insufficient understanding by the facilities staff of technical systems.

Stage 4 – Initial aftercare

Continuing involvement by the client, design and building team benefiting from lessons learned and the occupant satisfaction surveys that form part of the Soft Landings post-occupancy evaluation (POE) process. This period is intended to help occupiers understand their building and facilities staff to operate the systems. The aftercare team's workplace must be in a visible area and occupiers must be told of the purpose of their being there. They also need to undertake walkabouts to observe occupation and head off emerging problems.

Stage 5 – Extended aftercare and post-occupancy evaluation

This stage closes the loop between design expectation and the actual performance and involves periodic inspections over three years by the aftercare team to help users and operators to get the best out of the building. This may include fine-tuning systems to optimise energy efficiency and to take account of occupant feedback. In years two and three these become less frequent and include an occupant satisfaction survey which is used to make comparisons with other projects. It is recommended that a Soft Landings champion is appointed to the project team.

COMMISSIONING

Testing or commissioning?

It is important for the project manager to understand the differences between the terms testing, commissioning and performance testing, and to ensure that the programme has sufficient time within it to enable these activities to be undertaken. Unfortunately, with this stage of the project being so close to handover, there is often pressure to gain time by shortening the testing, commissioning and performance / environmental testing programme. This should be strongly resisted. Rarely, if ever, after the project will such an opportunity exist to fully test the services to ensure that they work individually, as a system, and under part-load and full-load conditions. Many problems with respect to the under-performance of services within an occupied building can be related back to either insufficient quality in the testing and commissioning or insufficient time to test and commission.

It should also be borne in mind that various statutory services will need to be demonstrated to site inspectors and insurers. Time should be allowed for within the programme since these activities are often taken as separate tests after the main commissioning has been undertaken.

Commissioning refers to the process of bringing an item into operation and ensuring that it is in good working order. On construction projects, this refers primarily to mechanical and electrical services.

Services requiring commissioning may include:

- heating, cooling and ventilation systems,
- back-up systems,
- telecoms,
- water supply and sanitation,
- fire detection and protection systems,
- information and communications technology systems,
- security systems,
- lifting equipment and escalators, and
- catering equipment.

The contract documents should set out:

- who will be responsible for commissioning different building services,
- what methods, standards and codes of practice are to be used,
- what should happen to test results, and
- whether commissioning is to be witnessed and if so, by whom.

Commissioning activities may include:

- ensuring client access and providing client training and demonstrations,
- completing operating and maintenance manuals, record drawings, software and test certification,
- obtaining statutory approvals and insurance approvals,
- manufacturers work testing,
- component testing,
- pre-commissioning tests,
- set to work – this is the process of switching on (i.e. setting to work) items such as fans and motors to ensure that they are operating as specified (for example, checking that fans are turning the right way),
- balancing – this follows setting to work and involves looking at whole systems (rather than individual components) to ensure that they are properly balanced (i.e. water is coming out of all the taps at the correct pressure, air is coming out of the correct diffusers etc.),
- commissioning checks and performance testing, and
- post-commissioning checks and fine-tuning during occupancy.

Testing

During services installation various tests will be undertaken known as 'static testing'. This testing is normally undertaken to prove the quality and workmanship of the installation. Such work is undertaken before a certificate is issued to 'enliven' (i.e. to make live) services, whether electrically or otherwise. Examples of this sort of testing are:

- pressure testing ductwork and pipework, and
- undertaking resistance checks on cabling.

Commissioning

Upon completion of static testing, dynamic testing can be undertaken – this is 'commissioning'. Commissioning is carried out to prove that the systems operate and perform to the design parameters and specification. This work is extensive and normally commences by issuing a certificate permitting the installation to be made 'live', i.e. electrical power on. After initial tests of phase rotation on the electrical installation and checking fan / pump rotation (in the correct direction), the more recognised commissioning activities of balancing, volume testing, load bank testing, etc. begin.

Performance testing

Upon completion of the commissioning, performance testing can begin. Some may not distinguish between commissioning and performance testing. However, for programming purposes it is worth distinguishing between commissioning plant as individual systems and undertaking tests of all plant systems together, known as performance testing, (and including environmental testing). Sometimes this performance testing is undertaken once the client has occupied the facility, e.g. for the first year because systems are dependent upon different weather conditions. In such cases, arrangements for contractor access after handover to fine-tune the services in response to changing demands must be made. However, for some facilities it may be necessary to simulate the various conditions expected to prove that the plant systems and controls operate prior to handover, e.g. computer rooms.

CLIENT COMMISSIONING

Having accepted the site from the contractor at practical completion, the client has to prepare the facilities for occupation.

The principles of client commissioning and occupation should be determined at the feasibility and strategy stage. The objective of client commissioning is to ensure that the facility is equipped and operating as planned. This entails the formation of an operating team early in the project so that requirements can be built into the contract specifications. Ideally, the operating team should be formed in time to participate in the design process.

It is common for the client to organise a separate project to carry out accommodation works. Often this team will be separate from the main project team and will comprise personnel with greater experience of operating in a finished project environment.

Typical elements of client accommodation works for an office building would be:

- Fitting out of special areas:
  - restaurant/dining areas,
  - reception areas,
  - training areas,
  - executive areas,
  - post rooms, and
  - vending areas.

- Installation of IT systems:
  - servers,
  - desktop PCs,
  - telecoms equipment,
  - fax machines, and
  - audio visual and video conferencing.
- Demountable office partitions:
  - furniture,
  - specialist equipment,
  - security systems, and
  - artwork and planting.

The main tasks of client commissioning include:

- Establishing the operating and occupation objectives in time, cost, quality and performance terms. Consideration must be given to the overall implications of phased commissioning and priorities defined for sectional completion, particular areas / services and security.
- Making sure that an appropriate allowance for client's commissioning costs is made in the budget. Accommodation works can account for as much as 3 per cent of the total construction budget.
- Arranging the appointment of the operating team.
- Preparing role and job descriptions (responsibilities, timescales, outputs) for each member of the operating team. These should be compatible with the construction programme and any other work demands on members of the operating team.
- Co-ordinating the preparation of a client's commissioning schedule and action list, using a commissioning checklist.
- Arranging appropriate access as necessary for the operating team and other client personnel during construction.
- Arranging co-ordination and liaison with the contractors and the consultants to plan and supervise services commissioning, e.g. preparation of new work practices manuals, staff training and recruitment of additional staff if necessary; the format of all commissioning records; renting equipment to meet short-term demands; overtime requirements to meet the procurement plan; meeting quality and performance standards and so on.
- Considering early appointment / secondment of a member of the client management team to act as the occupation co-ordinator; this ensures a smooth transition from a construction site to an effectively operated and properly maintained facility.

Before the new development can be occupied, the client needs to operationally commission various elements of the development. This involves setting to work various systems and preparing staff ready to run the development and its installations:

- transfer of technology,
- checking voice and data installation are operational,
- stocking and equipping areas such as restaurants,
- training staff for running various systems,
- training staff to run the property, and
- obtaining the necessary statutory approvals needed to occupy the building, such as the environmental health officer's approval of kitchen areas.

Occupation of the developed property is dependent on detailed planning of the many spaces to be used. For office buildings, this space planning process is developed progressively throughout the project life cycle. Final determination of seating layouts may be delayed until the occupation stage in order to accommodate the latent changes to the client's business structure. It is essential that for each of these stages client user panels have a direct involvement and approve each stage.

A typical space planning process consists of:

- confirming the client's space standards, including policy on open-plan and cellular offices,
- confirming the client's furniture standards,
- determining departmental headcount and specific requirements
- determining an organisational model of the client's business reflecting the operational dependencies and affinities,
- developing a building plan in order to fit the gross space of each department within the overall space of the building,
- developing departmental layouts to show how each department fits the space allocated to it, and
- developing furniture seating layouts in order to allocate individual names to desks.

Moving or combining businesses into new premises is a major operation for a client. During the duration of the moves there is potential for significant disruption to the client's business. The longer the move period, the greater the risks to the client. Migration therefore requires a significant level of planning. Often the client will appoint a manager separate from the new building project to take overall responsibility for the migration. For major or critical migrations, the client should consider the use of specialist migration consultants to support their in-house resource.

During the planning of the migration a number of key strategic issues need to be addressed. As some of these strategic issues could have an impact on the timing and sequencing of the main building works, it is important to address them early in the project life cycle:

- determining how the building will be occupied,
- establishing the timing of the moves,
- identifying the key activities involved in the migration and assigning responsible managers,
- determining move groups and sequence of moves to minimise business disruption,
- determining the project structure for managing the move,
- identifying potential risks that could impact on the moves, and
- involving and keeping the client's staff informed.

The final part of occupation is the actual move management. This involves the appointment of a removal contractor, planning the detailed tactics of the moves, and supervision of the moves themselves. The overall period during which the moves will be undertaken is determined by the amount of 'effects' to be transferred with each member of the staff and by the degree of difficulty of transferring IT systems for each move group.

A critical decision for the client during the occupation stage is the point at which a freeze is imposed on space planning and no further modifications are accommodated until after migration has been achieved. It is likely that the factor having most impact on the timing of the freeze date will be the setting up of individual voice and data system profiles. It would be common for clients to impose an embargo on changes both sides of the migration and for the client then to carry out a post-migration sub-project to introduce all the changes required by departments.

OPERATION AND MAINTENANCE

Several systems have been developed to aid the client to effectively manage the operation and maintenance of the completed project. The project manager should be aware of these systems – some of which can be described as work in progress.

COBie

Construction Operations Building Information Exchange (COBie) is a data format for the publication of a subset of building model information focused on delivering building information not geometric modeling. It is closely associated with BIM and was devised by Bill East of the United States Army Corps of Engineers in 2007. COBie is formal schema that helps organise information about new and existing facilities. It is general enough that it can be used to document both buildings and infrastructure assets. It is simple enough that it can be transmitted using a spreadsheet (Excel) format. It is a means of sharing structured information, just like BIM, except that it is not as comprehensive as a full BIM, but nevertheless it is a step in the right direction.

COBie helps capture and record important project data at the point of origin, including equipment lists, product data sheets, warranties, spare parts lists and preventive maintenance schedules. This information is essential to support operations, maintenance and asset management once the built asset is in service. In December 2011, it was approved by the US-based National Institute of Building Sciences as part of its National Building Information Model (NBIMS-US) standard. COBie has been incorporated into software for planning, design, construction, commissioning, operations, maintenance and asset management.

The information is stored on Excel spreadsheets; from the client's perspective he / she may ask for the delivery of COBie from the lead designer and / or lead contractor to support the timely delivery of information to support the management of the facility. A complete COBie should be expected at the time of handover, but earlier interim deliveries can be used to monitor the business case for the facility and to help plan for taking ownership.

The COBie information can be either be kept as delivered, or held in ordinary databases, or it can be loaded into existing facility management and operations applications, either automatically or using simple cutting and pasting. The client should be explicit about the purposes for which the information is required and about the timing and content of any interim deliveries.

From the designers' and contractors' perspective, COBie allows the team to record their knowledge about a facility in both its spatial and physical aspects. Spatially, it can document the spaces and their grouping into floors / sectors and into other zones. Physically, it documents the components and their grouping into product types and into other systems. Usually the information needed for COBie will be available already, either in BIM models or in reports and schedules and in other material prepared for handover.

For product specifiers and suppliers, COBie can be used to provide product data to support the specification / selection / replacement process. If the client requirements include this, then the product types should be given the specific attributes appropriate to that type. There are currently 700 templates available in a range of formats and are freely available (open access).

Alignment of NRM 3 to COBie II data structure and definitions for Building Information Modelling

Data should be made interoperable, for example through the COBie data exchange format, in order to ensure that building maintenance cost data is accessible for life cycle costing of construction projects, and to ensure that output data from life cycle cost of maintenance models is accessible to other interoperable models (for BIM cost modelling and setting up asset information systems to deliver maintenance programmes of works).

Where maintenance and renewal works unit rates are used for order of estimates and cost planning during the design and construction phases, the output from the cost analysis of post-construction maintenance works could be structured into a COBie format. This data can then be interoperable and enable the integration of life cycle cost of construction and maintenance works.

The building information maintenance model (whether generated during pre-construction or during post-construction, in use) should be provided in the same format and data referenced to physical assets or systems, types (specifications) and components – as well as linking to the building, blocks, zones, floors and spaces. (Note that mapping of the NRM 1 data structure to COBie is included in BS 8544;2013 guide for life cycle costing of maintenance during the in-use phases of building.)

Table 4.1 shows the COBie II data classifications used for Building Information Modelling (BIM). The main differences are in spatial and physical classifications (which is outside the scope of these rules).

NRM 3 elemental cost data classification aligns with the COBie II data classifications, notwithstanding minor differences in definitions stated below:

- element (or systems),
- component (or sub-element / systems)
- specification (which COBie calls type),
- tasks or actions required (which COBie calls job),
- resources
- spares (including materials and consumables), and
- other costs (user defined).

Table 4.1 Definitions from COBie II data classifications

Sheet	*Contents*
Facility	Includes the project, site and building / structure
Floor	Sectors are the mandatory spatial structure
Space	The spatial locations where inspection, maintenance and operation jobs occur
Zone	The mandatory grouping of components as types or products, used to organise maintenance tasks
System	Additional functional groupings of components
Component	The physical assets
Type	The mandatory grouping of components as types or products, used to organise maintenance tasks
Job	The processes used to maintain and operate the assets
Spare	The physical objects
Resources	Support the processes

Source: PAS 1192 Part 2.

Table 4.1 also highlights the importance of classifying the cost and asset information back to the relevant space, building, locational and functional data conventions, to create robust maintenance and renewal cost plans.

When the design and construction process requires a BIM model to be used, then it is essential to ensure the elemental cost plan is interoperable. Classifications of asset classes, or grouping of elemental or system types, may need to be applied to named objects to support BIM cost modelling option studies. More detailed guidance on BIM is provided in PAS 1192 Part 2. How this relates to life cycle costing of maintenance in use is provided in BS 8544 and other published sources listed in the Further reading section.

Note that COBie is a standardised tabular representation of a facility and its constituents allowing the exchange of their detailed properties and impacts such as maintenance cost and carbon. COBie is a subset of IFC schema.

PAS (Publicly Available Specifications) 1192–3

BSI released PAS 1192–3 *Specification for information management for the operational phase of construction projects using building information modelling* for public consultation which closed early in December 2013. This is a partner document to PAS 1192-2. While Part 2 focuses on the delivery phase of projects, this new document focuses on the operational phase of assets, being about the availability, integrity and transfer of data and information during this phase. The document specifies how information from the Project Information Model (PIM) is transferred to the Assets Information Model (AIM) or how an AIM is created for an existing asset. Of equal importance is how information is then retrieved and passed on to an existing enterprise system such as a database. While it is not explicit in what data is to be covered, it does cross-refer to broad headings and documents which will define data content.

Unlike Part 2, which follows a clear sequence through the project stages, Part 3 describes both a mixture of planned and unplanned events in the life of an asset that could happen in any order between the point of handover and disposal. PAS 1192–3 is intended for those responsible for the management of assets, including their operation, maintenance and strategic management. While facilities management has a distinction between hard FM and soft FM, PAS 1192–3 uses the terms 'Asset' and 'Asset management' to refer to physically related requirements. One of the key messages from the government is to 'produce the right information, at the right time, at the right level of detail and definition'.

PERFORMANCE MEASUREMENT

Benchmarking facilities

Performance measurement is an integral part of business management. By championing key company and project aims (key performance indicators [KPIs]), managers are more likely to achieve success. But the only way of knowing whether those goals are being delivered is by identifying indicators of their success and using them to determine the way the business is performing.

Benchmarking is not just about cost levels as there are variety of issues around facilities management that can be benchmarked.

Post-occupancy evaluation (POE) is the process of evaluating a development to determine:

- how successful its delivery was,
- how successful the completed development is,
- where there is potential for further improvement, and
- what lessons can be learned for future projects.

Designers are often guided by the constraints of a project as it unfolds, but continual learning and dissemination of acquired knowledge holds the key to shaping the future of projects and practices. The concept originally surfaced in the 1970s and the drive towards tighter environmental targets. Today, new regulations and a focus on a more sustainable approach is driving a resurgence in post-occupancy evaluation. It is central to improving the performance of low and zero carbon building design, vital for sustainable construction. Without post-occupancy evaluation, the sustainability of buildings in occupation cannot be properly understood.

The process of post-occupancy evaluation can be visualised as part of the building life cycle, where information learned from an operational (and occupied) project can be used to inform decisions at all of the stages in the design and operational life of a building. Post occupancy evaluation can be particularly valuable to repeat developers and may be a requirement of some funding bodies.

Post-occupancy evaluation may be carried out by a member of the consultant team, independent client advisers or by an in-house team established by the client. As post-occupancy evaluation is likely to take place after the main construction contract has been completed, the consultant team's involvement will have come to an end unless post-occupation services were a specific requirement of the original appointments. Ideally the client should commit to carrying out post-occupancy evaluation at the beginning of the project so that appointment agreements and briefing documents include requirements to test whether objectives were achieved.

Post-occupancy evaluation may comprise two studies:

- a post project review to evaluate the project delivery process, and
- an assessment of performance in use.

The origin of the term 'benchmarking' is in surveying and levelling and typical benchmark signs can be seen on kerbs and walls throughout the country. It has particular relevance to construction programme managers as well as project managers.

Benchmarking is a generic management technique that is used to compare performance between varieties of strategically important performance criteria. The Xerox Corporation in America is considered to be the pioneer of benchmarking. In the late 1970s, Xerox realised that it was on the verge of a crisis when Japanese companies were manufacturing photocopies cheaper than it cost Xerox to manufacture a similar product. Another strong advocate of benchmarking is the automotive industry who successfully employed the technique to reduce manufacturing faults. Benchmarking can be broadly divided as:

- international,
- competitive,
- functional, and
- generic.

However, perhaps a more useful distinction is:

- output benchmarking, and
- process benchmarking.

Benchmarking is all about improvement, and not merely justifying existing levels, or achieving the average of the peer group. Benchmarking can be used between different organisations or within a single organisation provided that the task being compared is a similar process. It is an external focus on internal activities, functions or operations aimed at achieving continuous improvement. Construction, because of the diversity of its processes and products, was one of the last industries to embrace objective performance measurement and is still regarded with scepticism by some.

By implementing benchmarking, a client can bring considerable improvements to operating and maintenance costs. For continuous improvement to occur it is necessary to have performance measures which check and monitor performance to verify changes and the impact of improvement initiatives to understand the variability of the process. In general, it is necessary to have objective information available in order to make effective decisions.

Through the implementation of performance measures (what to measure) and the selection of the measuring tools (how to measure), an organisation or a market sector communicates to the outside world and clients the priorities, objectives and values that the organisation or market sector aspires to.

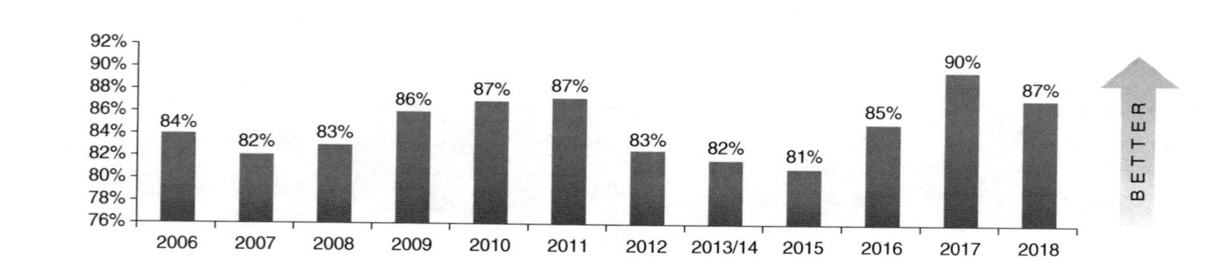

Figure 4.1 Client satisfaction

Source: CITB Industry Performance Report 2018.

Performance measurement

Performance measurement demonstrates whether an individual or company is achieving continuous improvement. Benchmarking provides a 'yardstick' by which to judge performance. Some sample KPIs are listed below:

- client satisfaction,
- defects,
- construction time and cost,
- productivity,
- profitability,
- health and safety,
- employee satisfaction,
- staff turnover,
- sickness absence,
- working hours,
- qualifications and skills,
- impact on environment,
- whole life performance,
- waste, and
- commercial vehicle movements.

The latest CITB survey of KPIs revealed that overall client satisfaction in 2018 was 87 per cent, as illustrated in Figure 4.1, and that client's satisfaction with the finished product was overall 8 out of 10, slightly down on the previous year.

5

Occupancy / RIBA Plan of Work Stage 7

Even when the project is complete and the client has taken possession of the built asset, the project manager still has an important role to play. For clients with large property portfolios, the project manager should explain the benefits of post-occupancy evaluation of the newly completed project as evaluation and benchmarking are the cornerstones of continuous improvement. Many buildings do not perform as planned; in some cases issues can impact on running costs, staff and client satisfaction and performance, as well as health and safety and comfort. For repeat construction clients, learning from and correcting past mistakes in design and commissioning of buildings can be extremely cost-effective and greatly improve workplace productivity for future projects. The tools that are available to the project manager at this stage are;

- project audit, and
- post-occupancy evaluation (POE), which includes such techniques as:
    - focus groups,
    - visual surveys,
    - energy use surveys, and
    - building walk-throughs.

PROJECT AUDIT

A project audit checks that everything in the project is running according to plan. As the title suggests, the process is very similar to any audit and the project manager shouldn't have anything to worry about regarding the process – it's a way of ensuring that the project stays on track.

A project audit may be carried out during the currency of the project or when the project is complete. If carried out while the project is still on-site, it enables the project manager, project sponsor and project team an interim view of what has gone well and what needs to be improved to successfully complete the project. If done at the close of a project, a project audit can be used to develop success criteria for future projects by providing a forensic review. This review will provide an opportunity to learn what elements of the project were successfully managed and which ones presented some challenges. This will help the organisation identify what it needs to do so that mistakes are not repeated in the future.

A project audit consists of three stages:

- Stage 1 – Establishment of success criteria and questionnaire development.
- Stage 2 – Background research.
- Stage 3 – Report development.

Stage 1 – establishment of success criteria and questionnaire development

Success criteria

Interview the core project sponsors, including the project manager, to determine their success criteria for the project audit. This ensures that their individual and collective needs are addressed.

Questionnaire development

Develop a questionnaire to be sent to each member of the core project team and selected stakeholders. It has been found that individuals will often complete a questionnaire in advance of an interview; it helps to focus their thoughts. The interview provides the facilitator with the opportunity to gain deeper understanding of the interviewee's comments. The questionnaire allows them to reflect on the project's successes, failures and challenges as well as missed opportunities.

Project audit questions

It is easiest to use open-ended questions for the interview. These questionnaires can be used for team members and / or stakeholders who cannot attend an interview. Frame the questions in such a way as to help team members identify the major project successes, issues, concerns and challenges, how the team worked together, how reporting meetings were handled and how risk and change were managed.

Questionnaires

Questionnaires can be a valuable and efficient way to collect data from a large group of people and can either be in a hard copy or, more likely, web-based. Good questionnaires require thought and can assure anonymity of the respondents. There are a few golden rules for questionnaire preparation (see Appendix E for an exemplar):

- Don't pose open-ended questions; remember the results have to be analysed.
- Keep the questions as short as possible; multiple pages with dozens of questions and asking for lengthy comments will put respondents off and the questionnaires will end up in the bin or deleted.
- Use multiple-choice questions with tick boxes.
- Use a large enough sample to produce meaningful results.
- Leave a box at the end for general comment.
- Pilot the questionnaire prior to going live.

To assist the process there are several online survey tools such as:

- SurveyMonkey – www.surveymonkey.com/
- PollDaddy – http://polldaddy.com/
- Zoomerang – www.zoomerang.com/
- Limesurvey – www.limesurvey.org/
- SurveyGizmo – www.surveygizmo.com/

The advantages of using an online survey tool are:

- simplicity,
- scalability,
- low cost (relative to other methods) – especially when there are a large number of questionnaires to process,
- reduced administrative burden,
- little or no (manual) data entry,
- reductions in error rates / cloudy data / messy data that accompany manual entry, and
- access to question banks, analytical tools and graphics engines.

Online survey tools generally come in two formats:

- free to access, and
- subscription format.

Free-to-access portals allow:

- the creation and publishing of surveys in an online / web-based setting,
- end-users to submit responses through a website (web-based forms, portals, etc.),
- the storage of the 'raw data' on a web-based server / database,
- the ability to export data to outside analytical programs such as Excel, etc., and
- the project manager to run queries, generate reports and create visual representations.

However, free-to-access portals do have limitations:

- only basic surveys are possible, with limited question types,
- usually there are limits on the size of survey and number of questions,
- in addition there are limits on size of response sample and number of responses,
- little or no skip logic,
- limited 'export' ability to Excel, etc.,
- limited data analysis, and
- limited capacity for graphics, visual displays, etc.

By contrast, subscription service online survey tools generally contain more features including;

- greater variety of question types,
- multiple modes of skip logic,
- access to question banks and survey banks,
- easy export to statistical packages such as Excel, etc.,
- sophisticated analytical tools, and
- the ability to generate graphics and visual displays of data.

When it comes to choosing the portal for the design and distribution of the questionnaires, the following considerations should be taken into account:

- compare costs – is free really the best option?
- test out multiple platforms, both front-end and back-end, and
- test out the support models for each platform – reliability, flexibility, accuracy, responsiveness, etc.

Before starting to draft a questionnaire, it is important to let the research and reporting needs drive the process in terms of:

- audience,
- sample size,
- question type (quantitative vs qualitative),
- statistical / analytical need,
- report format / models.

Stage 2 – background research

Background research should be carried out along the following lines:

1. Conduct individual interviews with two main groups to identify past and current issues, concerns, challenges and opportunities:
   (a) the project sponsor, project manager and project team members, and
   (b) stakeholders, including suppliers, contractors and other internal and external entities.
2. Assess the issues, challenges and concerns in depth to establish the root cause of any problems.
3. Review all historical and current data relating to the project including:
   (a) team structures,
   (b) scoping studies,
   (c) project plan,
   (d) meeting minutes,
   (e) actioned items,
   (f) risk registers, and
   (g) change registers.
4. Review the supply chain and its management.
5. Interview stakeholders to discover whether their expectations have been realised.
6. Review the Project Quality Management Plan (see Chapter 4).
7. Identify lessons learned for future projects.

Stage 3 – report development

The project audit should be prepared using the following steps:

- Collate the information collected from the interviews and the individuals who completed the questionnaire.
- Consolidate the findings from the project documentation review.
- From the above, identify the issues, concerns and challenges and highlight the potential opportunities.
- Identify lessons learned that can improve performance going forward.
- Draft the report and recommendations on the basis of the findings.
- Present the report.

Conclusion

The main purpose of a project audit is to highlight lessons learned that can help performance of future projects by undertaking a forensic review to uncover problems. In this way, project audits are very beneficial and can provide the following outcomes:

- development of lessons learned to be applied to future projects,
- development of project success factors,
- recognition of risk management so that risk assessment and the development of associated contingency plans become the norm, and
- development of criteria to improve relationships and communication in the supply chain.

POST-OCCUPANCY EVALUATION (POE)

Post-occupancy evaluation (POE) is the process of obtaining feedback on a building's performance in use. The value of POE is being increasingly recognised, and it is becoming mandatory on many public projects. POE is valuable in all construction sectors, especially healthcare, education, offices, commercial and housing, where poor building performance will impact on running costs, occupant wellbeing and business efficiency.

POE consist of three phases and is best carried out over several years:

- **An operational review** – carried out three to six months after occupation, this is a review of the process and the functional performance of the new building. The first of the POE reviews, an operational review considers such items as:

- **Process** – procurement, design, commissioning and occupation.
- **Function** – does the facility meet original business objectives and provide a comfortable and serviceable space with optimal operational and running costs?
- **Technical** – end users, FM, consultants, contractors.

- **A project review** – carried out twelve to eighteen months after occupation. This review takes place a year or so after occupation, allowing a full seasonal cycle of information to be taken onto account. The project review seeks to carry out an in-depth review of the technical and functional performance of the project and identifies whether any adjustments are needed to the building or the building systems. Generally, performance-in-use assessments cannot begin until six to twelve months after occupation, as operations may not be properly established and the building will not have operated in all seasons. They may then be part of a continuous process. An assessment of performance in use can include:
    - **Business objectives:**
        - the achievement of business case objectives,
        - whole life costs and benefits against those forecast (including assessment of capital vs running costs),
        - whether the project continues to comply with the business strategy,
        - whether operations have improved,
        - the resilience of the development and business to change, and
        - business and user satisfaction (including staff and user retention and motivation).
    - **Design evaluation:**
        - the effectiveness of the space planning,
        - aesthetic quality,
        - the standards of lighting, acoustic environment, ventilation, temperature and humidity,
        - air-pollution and air quality,
        - user comfort,
        - maintenance and occupancy costs,
        - defects,
        - the balance between capital and running costs,
        - an assessment of whether the development is being operated as designed, and

- environmental and energy consumption in use. (Note that regular evaluation of energy consumption is mandatory for certain types of buildings under the Energy Performance of Buildings (Certificates and Inspections) (England and Wales) Regulations.)

- **Process** – this aspect of the review / evaluation should concentrate on how the project was delivered, for example:
  - the appointment of the design team and the development of the brief. Linking the brief to the client's strategic goals should also be included, and
  - the appropriateness of the procurement and contract strategy, etc.
- **Assessment** – the assessment should compare findings to the original targets set out in business case (the original targets may need to be updated to reflect changes to the project brief during the design process, inflation etc.). It should also compare findings to other projects and industry standards and compare the outcome of the project with the position had the project not taken place. A report should be prepared that identifies issues, recommends remedies, and makes suggestions for improvements in performance for future projects.
- **Other services** – could be provided by consultants during this period and might include providing advice on:
  - letting,
  - rating,
  - maintenance,
  - energy consumption,
  - insurance,
  - tenants' queries,
  - facilities management,
  - energy performance certificates,
  - BREEAM assessments, and
  - tender documents for maintenance and operation contracts.

- **Strategic review** – carried out three to five years after occupation. This review should take place several years after occupation in order to ascertain how the building format is likely to meet the future needs of the client / sponsor. Items covered in the previous reviews can be re-evaluated during this process. This will determine how the building format fulfils the client / sponsor's future needs.

It is important to remember that the processes of evaluation and benchmarking are the cornerstones of continuous improvement. Many buildings do not perform as planned – in some cases this can impact on running costs, staff and client satisfaction and performance, health, safety and comfort. For repeat construction clients, learning from and correcting past mistakes in design and commissioning of buildings can be extremely cost-effective and greatly improve workplace productivity.

The POE can cover a wide range of activities including:

- the process,
- the functional performance, and
- technical performance.

It should include both a qualitative and analytical assessment.

Post-occupancy evaluation will:

- highlight any immediate teething problems that can be addressed and solved,
- identify any gaps in communication and understanding that impact on the building operation,
- provide lessons that can be used to improve design and procurement on future projects, and
- act as a benchmarking aid to compare across projects and over time.

Post-occupancy evaluation methods can be tailored to individual needs, including:

- occupant and client consultation,
- environmental comfort and control over environmental conditions,
- building impact on productivity and performance, staff and user retention and motivation,
- customer experience and user satisfaction with amenities, image and layout,
- review of design, procurement, construction and handover processes,
- monitoring of environmental conditions – including temperature, noise, light, air quality, ventilation and relative humidity,
- assessment of design quality using BRE's DQM (Design Quality Monitoring) – a structured method for assessing design quality and building performance against industry benchmarks and good practice, and

- sustainability and utility audits – to measure and demonstrate the environmental performance of buildings in use, to inform property management and energy efficiency strategies.

A suggested approach

Identify the general need for the study and identify the need for any specialist consultants or advisors.

A suggested approach to carrying out a POE is outlined below:

1. Identify the POE strategy and the need for the evaluation. At this stage the consultants carrying out the study should be identified and whether the consultants will be internal or external. Decide the parameters of the study including:
   - what items are to be included in the study,
   - when the study is to be carried out,
   - the possible methodologies,
   - whether the results are to be benchmarked against other projects,
   - the format of the final report, and
   - whether the study will be carried out internally or with the use of external consultants.
2. Define the objectives and priorities. The timing should be identified and also whether it will be an in depth or superficial study.
3. Brief stakeholders. The stakeholders should now be briefed in the following ways:
   - hold a workshop / briefing meeting,
   - define timing and those involved, and
   - explain the methodology (questionnaires, workshops, interviews, etc.), including:
     - the objectives,
     - the timing,
     - who will carry out the study,
     - who should be involved,
     - specific issues that should be addressed, and
     - where the study / interviews will take place, etc.
4. Select methodology:
   - prepare materials for study (questionnaires, etc.),
   - prepare meeting schedules and agendas, and
   - agree the form of feedback.

5. Carry out POE:
   - depending on chosen methodology, arrange for the distribution of POE materials,
   - gather and collate study data, and
   - analyse data.
6. Prepare report:
   - decide on the format of the report and circulation, and
   - ask for comments on the draft.
7. Feedback / action:
   - finalise report, print and publish,
   - draw-up action list, and
   - monitor the process.

Methodologies

A number of methodologies are available and these include both quantitative and analytical approaches, for example:

- questionnaires,
- interviews,
- focus groups,
- visual surveys,
- building walk-throughs,
- energy use surveys, and
- interviews.

Questionnaires

See previous notes under Project audit section above.

Interviews

An alternative or an addition to questionnaires are interviews, the advantage of interviews over questionnaires are:

- interviews generally takes hours rather than days to complete,
- they are more focused and targeted than other approaches,
- they are easy to arrange, and
- they contain greater detail than some other methods.

Focus groups

Focus groups have proved to be a highly successful research technique for engaging a group of people with a question, product or building performance. Bringing together a group to discuss a particular topic can provide a more natural setting than one-to-one interviews, as it allows participants to share their experiences and, through discussion, can enable new strands of thought to emerge. Therefore, this qualitative research method can generate useful data in a less resource intensive manner than interviewing. Using a focus group to engage with questions of performance can form part of the design process of a wider survey, or it can uncover the opinions of key stakeholders. A focus group can be a useful addition to a questionnaire-based survey.

Visual surveys

With tighter budgets and even tougher time constraints, getting the right data is essential to empower the right decisions. Visual surveys still provide vital information to provide robust data in planning maintenance within an asset management framework. Visual surveys traditionally consist of sets of photographs of buildings, although increasingly include data collected by drones.

Energy use surveys (assessments)

As the name suggests, this technique involves determining the amount of energy being consumed by a number of sources. Probably best carried out by specialist, an energy use survey can include such items as:

- CO_2 emissions,
- water consumption,
- lighting,
- heating and ventilating, and
- insulation.

Building walk-throughs

This approach involves walking through the building with a pre-prepared observation sheet and can include a subjective narrative, on a room by room basis, of elements such as finishes, doors and windows, lighting, furniture and environmental quality.

Feedback

POE is a way of providing feedback throughout a building's life cycle from initial concept through to occupation. The information from feedback can be used for informing future projects, whether it is on the process of delivery or technical performance of the building. It serves several purposes, discussed below.

Short-term benefits

- Identification of and finding solutions to problems in buildings.
- Response to user needs.
- Improved space utilisation based on feedback from use.
- Understanding of implications of change on buildings, whether budget cuts or working context.
- Informed decision-making.

Medium-term benefits

- Built-in capacity for building adaptation to organisational change and growth.
- Finding new uses for buildings.
- Accountability for building performance by designers.

Longer-term benefits

- Long-term improvements in building performance.
- Improvement in design quality.
- Strategic review.

The greatest benefits from POEs come when the information is made available to as wide an audience as possible. Information from POEs can provide not only insights into problem resolution but also provide useful benchmark data with which other projects can be compared. This shared learning resource provides the opportunity for improving the effectiveness of building procurement, where each institution has access to knowledge gained from many more building projects than it would ever complete.

The Usable Buildings Trust

The Usable Buildings Trust (UBT) is an organisation dedicated to achieving buildings with better all-round performance through the effective use of feedback at all stages in their life cycles – not just in their initial construction, but in all aspects of feasibility, briefing, design, commissioning, occupation, use, management and adaptation. Although based in the UK, UBT has an international perspective. More details can be found at www.usablebuildings.co.uk. UBT Trustees, in alphabetical order, are Denise Bennetts, Roderic Bunn, Sir Andrew Derbyshire, Joanna Eley, John Field and Jim Meikle.

UBT has four main areas of activity:

1 **Research** – UBT promotes and liaises with research into understanding and improving building performance and undertakes its own research where necessary.
2 **Development** – new ideas or research outcomes often need development before they can be used effectively in practice.
3 **Networks and capacity building** – UBT works with individuals, organisations and networks who are, or are planning to, put feedback principles into practice and exchange information.

Unfortunately from 2014 to 2018, key UBT members suffered major health problems, making little new research possible. Work on the Landlord's Energy Rating continued, particularly on the potential for a Commitment Agreement protocol to eliminate the 'performance gap' between design predictions and in-use outcomes for new office buildings and major refurbishments, following the success of this approach in Australia.

Appendix A

Financial statement

New Office Block, Leeds

Financial Statement No. 9

Date 2nd April 2020

	£	£
Contract sum		8,474,316
Less Risk allowance		36,000
		8,438,316
Adjust for:		
Variation orders No's 1–48	25,000	
Provisional sums	12,600	
Projected variations	3,000	
Agreed contractor's claim	5,700	43,600
Anticipated final account		£8,481,916

Exclusions:

VAT
Professional fees

Appendix B

Practical completion certificate pro-forma for NEC4 form of contract

To: *(The Contractor)*	To: *(The Employer)*
Address:	Address:
Telephone:	Telephone:
Fax:	Fax:
Attention:	Attention:
Contract no:	
Contract title:	

Insert appropriate wording depending upon which form of contract is utilised. Modify accordingly when the certificate is issued in respect of a portion of the works in the case of partial or sectional completion.

	Day	Month	Year
Completion achieved on:			
The completion date is:			
The defects date is:			
The defects on the attached schedule are to be corrected within the *defects correction period* which ends on:			

Works checked by the *Supervisor*:

....
Signature: *Name:* *Date:*

Certified by the *Project Manager*:

....
Signature: *Name:* *Date:*

Appendix C

Final account pro-forma

STATEMENT OF FINAL ACCOUNT

for

REFURBISHMENT OF TENANTS' MEETING HALL TONBRIDGE

Architect
Gardiner & Partners
6 Derby Walk
Tonbridge TN4 8HN

Borough Technical Services
P.S. Brookes FRICS
Technical Services Group
67 Uxbridge Road
Tonbridge TN5 6JK

Contractor
J. Harris & Co. Ltd
37 Newton Terrace
Tonbridge TN3 8GH

26 June 2020

I / We the undersigned hereby certify that the gross total value of the final account for this contract has been agreed in the sum of £2,645,363.78

Two million six hundred and forty five thousand three hundred and sixty three pounds and seventy eight pence.

and that payment of this gross amount shall be in full and final settlement of this account, subject to any adjustments required following the Local Authority's audit and liquidated and ascertained damages which the employer may deduct and that I / we have no further claims on this contract.

Signed..
For and on behalf of ....
....
....
....

Date..................

FINAL ACCOUNT SUMMARY
for
REFURBISHMENT OF TENANTS' MEETING HALL TONBRIDGE

	Omissions £	Additions £	£
Contract sum			2,670,000.00
Less Contingencies			15,000.00
			2,655,000.00
From prime cost sums summary	18,325.00	16,899.00	
From provisional sums summary	13,300.00	2,689.00	
From provisional items summary	3,191.44	61.70	
From variation account summary	75,839.04	73,672.78	
Fluctuations		896.78	
Agreed claim		6,800.00	
	110,655.48	101,019.26	
	101,019.26		
	9,636.22		9,636.22
			2,645,363.78
Less amount paid in interim certificates nos 1–12			2,642,876.00
Balance due			**2,487.78**

Therefore in this example the sum of £2,487.78 is due for payment to the contractor in full and final settlement and the statement of final account can be signed.

Appendix D

Design / construction project sample risk list

Construction risks
Unidentified utility impacts
Unexpected archaeological findings
Changes during construction not in contract
Unidentified hazardous waste
Site is unsafe for workers
Delays due to traffic management and road closures
Design risks
Incomplete quantity estimates
Insufficient design analysis
Complex hydraulic features
Surveys incomplete
Inaccurate assumptions during the planning phase
Environmental risks
Unanticipated noise impacts
Unanticipated contamination
Unanticipated barriers to wildlife
Unforeseen air quality issues
External risks
Project not fully funded
Politically driven accelerated schedule

Public agency actions cause unexpected delays
Public objections
Inflation and other market forces
Organisational risks
Resource conflicts with other projects
Inexperienced staff assigned to project
Lack of specialised staff
Approval and decision processes cause delays
Priorities change on existing programs
Project management risks
Inadequate project scoping and scope creep
Consultant and contractor delays
Estimating and / or scheduling errors
Lack of co-ordination and communication
Unforeseen agreements required
Right of way (ROW) risks
Unanticipated escalation in ROW values
Additional ROW may be needed
Acquisition of ROW may take longer than anticipated
Discovery of hazardous waste during the ROW phase

Appendix E

Sample questionnaire format

Please respond to the following questions by either circling the appropriate number or by writing your answer in the space provided. All information will be treated in confidence.

Where 1 = poor and 5 = excellent

1. Design evaluation. How do you rate the following:

a) The effectiveness of the space planning? 1 2 3 4 5

b) Aesthetic quality? 1 2 3 4 5

For what reason? ....

c) Standards of lighting? 1 2 3 4 5

d) Standards of ventilation? 1 2 3 4 5

e) Levels of temperature? 1 2 3 4 5

f) Comfort? 1 2 3 4 5

g) ITC? 1 2 3 4 5

h) Reliability of facilities? 1 2 3 4 5

2. Business objectives

Further reading

APM (2018) *APM Body of Knowledge,* 7th edition, APM.
CIOB (2014) *Code of Practice for Project Management for Construction and Development*, 5th edition, Blackwell Publishing.
CIC (2007) *The CIC Scope of Services Handbook*, 1st Edition, RIBA.
Cooke, B. & Williams, P. (2009) *Construction Planning, Programming and Control*, 3rd edition, Wiley Blackwell.
Egan, J. (1998) *Rethinking Construction*, The Construction Taskforce.
El-Sabaa, S. (2001) The skills and career path of an effective project manager, *International Journal of Project Management*, 19, 1–7.
Fewings, P. (2019) *Construction Project Management: An Integrated Approach,* 3rd edition, Routledge.
Fondahl, J.W. (1961) *A Non-computer Approach to the Critical Path Method for the Construction Industry*, Dept. of Civil Engineering, Stanford University.
Harris, F., McCaffer, R., & Fotwe, F. (2013) *Modern Construction Management*, 7th edition, Wiley Blackwell.
Latham, M. (1994) *Constructing the Team*, HMSO.
Maslow, A (1954) *Motivation and Personality*, Harper & Row Publishers Inc.
Ministry of Defence (1999) *Building Down Barriers*, Department of Trade and Industry.
Moore, G. & Robson A. (2002) The UK supermarket industry: an analysis of corporate and social responsibility, *Business Ethics: A European Review*, 11, 25–39.
Morris, P.W.G. (1997), *The Management of Projects*, Thomas Telford.
Newton, R. (2016), *Project Management Step by Step*, 2nd edition, Pearson Business.
OGC (2009) *Managing Successful Projects with PRINCE2*, OGC.
Project Management Institute (2017) *A Guide to the Project Management Body of Knowledge (PMBOK)*, 6th edition, PMI.

RIBA (2013) *RIBA Plan of Work 2013 – Overview*, RIBA.
RICS (2003) *Learning From Other Industries*, RICS Project Management Professional Group.
RICS (2007) *Project Monitoring*, 1st edition, RICS guidance note.
RICS (2009) *Development Management*, 1st edition, RICS guidance note.
Sommerville, J. & Campbell, C. (2000) *Project Management: An Evaluation of the Client and Provider Attribute Paradigms.* 17th Annual ARCOM.
Thomas, D.B. & Miner, R.G. (2007) Building Information Modeling – BIM: Contractual risks are changing with technology, 23 November, *a/e ProNet*. www.aepronet.org/ge/no35.html
Young, B.A. & Duff, A.R. (1990) Construction management: Skills and knowledge within a career structure, *Building Research and Practice*, 18 (3), 183–192.

Glossary

Artificial intelligence (AI)

Artificial intelligence is a term for describing when a machine mimics human cognitive functions, like problem solving, pattern recognition and learning.

Asset management

Systematic and co-ordinated activities and practices through which an organisation optimally and sustainably manages its assets and asset systems, performance, risks and expenditures over their life cycles for the purpose of achieving its organisational strategic plan.

Benchmarking

The objective of benchmarking is to understand and evaluate the current position of a business or organisation in relation to best practice and to identify areas and means of performance improvement.

Blockchain

A blockchain platform records transactions, agreements and contracts across a peer-to-peer network of computers worldwide.

BREEAM

BREEAM is the Building Research Establishment's Environmental Assessment Method and rating system for buildings and sets the standard for best practice in sustainable building design, construction and operation.

Building Information Modelling (BIM)

Building Information Modelling (BIM) is an intelligent model-based process that provides insight for creating and managing building and infrastructure projects and includes solutions for design, visualisation, simulation, quantification, facilities management and collaboration.

Change management

Change management is a systematic approach to dealing with change, both from the perspective of an organisation and on the individual level to achieve a required business outcome.

CIC Scope of Services

Multi-disciplinary scope of services published by the Construction Industry Council (CIC) for use by members of the project team on major projects.

Commissioning

The process of verifying that a new building or facility's sub-systems (for example plumbing, electrical and lighting, heating, ventilation and air conditioning, life safety, wastewater, controls and security) achieve the project requirements as intended by the building owner and as designed by the building architects and engineers.

Common data environment (CDE)

Single source of information for any given project, used to collect, manage and disseminate all relevant approved project documents for multi-disciplinary teams in a managed process.

Construction Operations Building Information Exchange (COBie)

COBie is a standardised tabular representation of a facility and its constituents allowing the exchange of their detailed properties and impacts such as maintenance cost and carbon. COBie is a subset of IFC schema.

Contractor designed portion (CDP)

Contractor designed portion refers to an agreement for the contractor to design specific parts of the works. The contractor may in turn subcontract this design work to specialist subcontractors. CDP should not be confused with design and build contracts where the contractor is appointed to design the whole of the works.

The Construction (Design and Management) Regulations 2015 (CDM)

The Construction (Design and Management) Regulations 2015 (CDM) controls site work and has health and safety responsibilities, including checking working conditions are healthy and safe before work begins, and ensuring that the proposed work is not going to put others at risk.

Environmental Impact Assessment (EIA)

The EIA Directive 2011/92/EU of the European Parliament and the Council of 13 December 2011 effects certain public and private projects on the environment and requires that an environmental assessment to be carried out for certain projects which are likely to have significant effects on the environment by virtue of their nature, size or location, before development consent is given.

Facilities management

An interdisciplinary process focusing on the long-term maintenance and care of buildings and facilities to ensure their functionality and support for their primary activities.

Framework agreement

Framework agreements are agreements with one or more suppliers which set out terms and conditions for subsequent procurements.

Industry Foundation Class (IFC)

The Industry Foundation Classes (IFC) data model developed by building SMART is an open, international and standardised specification for Building Information Modelling (BIM) data that is exchanged and shared among software applications used by the various participants in a building, construction or facilities management project.

Key performance indicators (KPIs)

Performance measurement for the construction industry that demonstrates continuous improvement over a range of predetermined metrics.

Lean

Production focused on delivering value for the employer or client and eliminating all non-value-adding activities using an efficient workflow.

OGC Gateway

The OGC Gateway Review process offers a structure for public sector projects based around a series of independent peer reviews carried out at key stages to verify that projects should be allowed to progress to the next stage. Archived on 22 August 2011 but still widely referred to.

Partnering

Partnering is a co-operative / collaborative relationship between business partners, formed in order to improve performance in the delivery of projects.

Partnering may be considered as a set of collaborative processes which emphasise the importance of common goals.

Practical completion
The stage at which the client is able to take possession of and occupy a project even though the building work may not be completed finished.

PRINCE2
PRINCE2 or PRojects IN a Controlled Environment is a project methodology developed by the private sector and adapted for use in the public sector originally for use on IT projects. The system is not a software package but can be used on a range of projects from small individual ones to mega projects.

Publically Available Specification (PAS) 1192-2
PAS 1192-2 provides specific guidance for the information management requirements associated with projects delivered using BIM. Not all information on a project will be originated, exchanged or managed in a BIM format. The intended audience for this PAS includes organisations and individuals responsible for the procurement, design, construction, delivery, operation and maintenance of buildings and infrastructure assets.

Soft Landings / Government Soft Landings
Soft Landings is a strategy adopted to ensure the transition from construction to occupation is as seamless as possible and that operational performance is optimised.

Sponsor
A person or organisation that provides support for a project and importantly takes responsibility for, among other things, funding.

Stakeholder
A person, group or organisation that has interest or concern in a project. Stakeholders can affect or be affected by the organisation's actions, objectives and policies. Some examples of key stakeholders are creditors, directors, shareholders, suppliers, unions and the community from which the business draws its resources.

Supply chain
The sequence of processes involved in the complete manufacture and distribution cycle – this could include everything from design through materials

and component ordering, through manufacturing and assembly, until the finished product is in the hands of the final owner.

Value engineering

Value engineering is based on a methodology developed by Lawrence D. Miles, who worked for the General Electric Company after the Second World War. It is a function-orientated technique that generates alternative ways to deliver a required outcome.

Value management

Value management involves emphasis on problem solving as well as exploring in depth functional analysis and the relationship between function and cost and a broader appreciation of the connection between a client's corporate strategy and the strategic management of the project.

Index

Note: Page numbers in *italics* refer to figures; those in **bold** refer to tables. Glossary entries are indicated by the letter 'g' (e.g. 275g).

@RISK 140–141
2D CAD 62–64
3D printing 66
40-hour value engineering methodology 99, *100*

acceleration 198–201
accommodation works 236–237
accountability 15, 130–131
adjudication 196
advance payments 197, 214
aftercare 230–231, 233
Ain Ahlia Insurance Company (AAAI) 18
alliancing 154
Anti-terrorism, Crime and Security Act 2001 17
APM (Association for Project Management) 10–11
Appointing a Project Manager 84
architects 76, 101–102, 115, 181
arrow diagramming 8
artificial intelligence (AI) 66–67, 272g
asbestos 120–121
assertiveness 33
asset management 272g
Assets Information Model (AIM) 243
Association for Project Management (APM) 10–11
Athos, Anthony 38
auctions 164
autocratic leadership 27

bankruptcy 201–204
belongingness **31**
benchmarking 5–6, 243–247, 272g
big bang migrations 226
BIM *see* Building Information Modelling (BIM)
BIM manager / co-ordinator 77–78
blockchain technology 67, 272g
bonds 214–215
BREEAM (Building Research Establishment's Environmental Assessment Method) 92, 93–95, *95*, 272g
Brexit 156–157
bribery 14, 17–18
BS 8544 242
BS ISO 15686–5 95
BSRIA (Building Services Research and Information Association) 84–85, **86–87**, 230; *see also* Soft Landings
budgetary control 35–36
building control 82, 225

Building Cost Information Service Standard Form of Cost Analysis (SFCA) 173, 175
Building Information Modelling (BIM): overview 57–66, *58*, **63**, 272g; interoperability 242; outputs and data drops 102–104, 232–233; practical completion 223; pre-construction phase 89–92
The Building Management Notebook 12
Building Regulations 92, 94, 151
Building Research Establishment's Environmental Assessment Method (BREEAM) 92, 93–95, *95*, 272g
building walk-throughs 259
business cases 19–20, 87–89
business objectives 254
business process re-engineering 40–41

CAD 61–64
Calgary Winter Olympics (1988) 9
cash flow 194, 196–198, 205
CDE (Common Data Environment) 77, 273g
CDM *see* The Construction (Design and Management) Regulations 2015 (CDM 2015)
CDP (contractor designed portion) 106–107, 150–151, 273g
Certificate of Practical Completion 222
change management 36–40, *37*, 273g
Channel Tunnel project 9
Chartered Institute of Builders (CIOB) 12
CIC Scope of Services 70, **71**, 273g
civil engineer 80
clerk of works 181
client commissioning 236–239
client satisfaction *246*, 247
clients: BIM and 62; CDM responsibilities 114–115, **117**; expectations of 45; roles of 74, **75**, 179–181; taking possession 223–225
closing projects 52
cloud-based software applications 66
COBie (Construction Operations Building Information Exchange) 62, 108, 240–242, **242**, 273g
Code for Measurement Practice 171
Code for Sustainable Homes 96
Code of Practice for Project Management for Construction and Development 12
codes of ethics 20–21
collateral warranties 215–217
commissioning: client commissioning 236–239; definition 273g; performance testing 236; or testing 234–235
Common Data Environment (CDE) 77, 273g
communication 31–35
competition 145
competitive advantage 19
competitive dialogue 160–161
competitive procedure with negotiation 161–162
completion: partial 221; practical 218–221; sectional 221
concept design 102
concessions 163, 168
confidentiality 16
conflicts of interest 16
confrontations 33–34
constraints 22–24, *23*
The Construction (Design and Management) Regulations 2015 (CDM 2015) 81, 113–120, **117–119**, 203, 273g
construction management 149
construction managers 80–81
Construction Operations Building Information Exchange (COBie) 62, 108, 240–242, **242**, 273g

construction phase: acceleration 198–201; BIM and 104; clients' role **75**; Construction Phase Plan 116; cost control / financial reporting 193–198; environmental management systems (EMS) 186–189; insolvency 201–204; insurances 212–217; modern methods of construction (MMC) 184–186; project team members 179–182; quality management 182–184; quantity surveyors 79; supply chains 205–212; works on-site 189–193
Construction Project Management (CPM) 8
construction project managers: digital construction and 57–67; role of 52–56
contingencies 176
contingency plans 131
contract administration *180*
contract administrator 76–77
contract award notices 163
contract management 149
contract notices 163
contractor designed portion (CDP) 106–107, 150–151, 273g
contractors 81, 116, **118**, 151–152, 182
contracts 150–151
Contracts (Rights of Third Parties) Act 1999 216
Control of Asbestos Regulations (CAR) 2012 120–121
conversation 32–35
Cooke, Robert 18
corrupt practices 14–18
cost advice 89, 121–129, 170–178
cost–benefit analysis (CBA) 142
cost constraints 23
cost consultants 78–79
cost control 193–198
cost overruns 45
cost reimbursement contracts 150
cost-plus contracts 150
Critical Path Method (CPM) 8, 50
culture 19

dangerous substances 120–121
data: COBie 240–242, **242**; migrations 225–227; ownership of 65
decision trees 139–140, *140*
default notices 195
defects 221–222
delegative leadership 28
Deming, William Edwards 6–7
design 99–107, *105*, 229
design and build (D&B) 146–148, 170
design and manage 149–150
design development 78, 103, 233
design evaluation 254–255
design phase, clients' role **75**
design responsibility matrix **107**
designers 76, 101–102, 115, **117–118**
development manager 3–4
digital construction 57–67
Disability Discrimination Act 2005 112–113
drones 66

East, Bill 240
EcoHomes points 96
electronic auctions 164
employer's agent 3
energy use surveys 259
Enforcement Directives 157
enhanced design and build 147
Enterprise Zones 109
environmental health officers (EHO) 81–82
Environmental Impact Assessment (EIA) 165–166, 274g
environmental issues 92–96

environmental management systems (EMS) 186–189
Equality Act 2010 112–113
esteem needs **31**
'An ethical framework for the global property market' 15–16
ethics: the business case 19–20; concepts of 12–13; decision-making models 20–21; the law and 16–18; the project manager and 13–16; unethical behaviours, activities and policies 18
EU public procurement 156–165

facilities management 62, 224–225, 274g
feasibility studies 88–89
feedback 52, 231, 260
final account 222, 265–266
financial reporting 193–198
financial responsibility 16
financial statements 193–194, 262
fire prevention 213–214
fire safety inspectors 82
first tier partners 210–211
fishbone diagrams 137–138, *137*
focus groups 259
Fondahl, J.W. 9
forward renewal works 127
framework agreements 155–156, 167, 274g
fraud 14
fuel supplies 225
full planning permission 108

Gantt, Henry 6
Gantt charts 6, 49–50, **51**
general review meetings 192
Government Procurement Agreements (GPA) 165
Government Soft Landings (GSL) 84–85, 230–232, 275g
green issues 92–96
gross external area (GEA) 170
gross internal area (GIFA) 171
guarantees 215–217

Hammer, Michael 40–41
handover 104, 218–223, 227–228
health and safety: Construction (Design and Management) Regulations 2015 (CDM 2015) 113–116, **117–119**; Construction Phase Plan 116; dangerous substances 120–121; Health & Safety Plan 116–120
Housing Grants, Construction and Regeneration Act 1996 194, 202, 204
human skills 25, *26*
Humphrey, Albert **47**

implementing projects 52
inception and briefing 233
Industry Foundation Classes (IFC) 274g
information managers 77
information seeking 34
initial risk registers 176
initiating projects 46–48
injury 212
innovative partnerships 162
insolvency 201–204
Institute of Business Ethics (IBE) 13, 20–21
insurances 65, 84, 85, 212–217, 225
integrity 16
intellectual property rights 65
interim payments 194, 197, 204–205
International Ethics Standards Coalition (IESC) 15–16
International Property Measurement Standards (IPMS) 171–172
International Society of American Value Engineers (SAVE) 96, *98*, *100*
interviews 258
in-use phase **75**
investment 19

Ishikawa (fishbone) diagram 137–138, *137*
ISO 9000 188
ISO 14000 series 188–189
ISO 21500 44

JCT (16) 53, 151, 212–214, 216
JCT construction contracts 150–151
JCT Design and Build (16) 216
Joint Fire Code 213–214
Just in Time (JIT) 212

key performance indicators (KPIs) 247, 274g
Kotter, John 39–40
Kotter's 8 Step Change Model 39–40

latent defects 221–222
law and ethics 16–18, 20
lawfulness 16
lead designers 76, 101–102, 115, **117–118**
leadership 26–29, *26*
Lean 210, 274g
LEED (Leadership in Energy and Environmental Design) 92, 95–96
legacy 20
legislation: Building Regulations 110–111; disability legislation 112–113; Enterprise Zones 109; environmental 92–93; ethics and 16–18, 20; health and safety 113–121; party wall issues 111–112; planning permission 107–109; rights of light 112
Lewin, Kurt 38
Lewin's Change Management Model 38
licensing of BIM 65
light, rights of 112
Local Democracy, Economic Development and Construction Act 2009 194–196, 202
Local Partnerships 168
local planning authorities (LPAs) 82, 92
Localism Act 2011 108

main contractors 81, 115–116, **118–119**, 182
maintenance 125–129, 239–243
management, points of 7
management contracting 148–149, 170
management procurement 148–150
Maslow, Abraham 30, *30*, **31**
McKinsey 7-S Model 38
mechanical and electrical (M&E) services 102, 184
meetings 32, 189–193
Microsoft Project 44
migrations: data 225–227; strategy 223, 227–228, 238–239
Miles, Lawrence D. 97
mitigation 131–132
modern methods of construction (MMC) 184–186, 229–230
monitoring and control 52, 76–77
Monte Carlo simulation 140–141
morale 19
most economically advantageous tender (MEAT) 158–159
motivation 29–31
move management 239
multiple projects 67–69

National Building Information Model (NBIMS-US) 240
NEC4 (New Engineering Contract) 54–56, 201, 213, 214, 263–264
needs 29, 30–31, *30*, **31**
negotiated contracts 153
negotiated procedures 160
negotiation 144–145
net internal area (NIA) 171–172
The New Economics for Industry, Government, Education 6–7

New Engineering Contract *see* NEC4 (New Engineering Contract)
New Models of Construction Procurement (NMCP) 156
Newforma 44–45
novated design and build 148
NRM 1 (RICS New Rules of Measurement 1) 89, 172–178
NRM 3 (RICS New Rules of Measurement 3) 89, 125–129, 241–242

occupancy phase: definition 218; post-occupancy evaluation (POE) 253–261; project audit 248–253
off-site construction 184–186, 214, 229–230
off-site manufacture (OSM) 184–185
OGC Gateway: overview 70, **71**, 274g; order of cost estimating *174*; Stages 0–3C *see* pre-construction phase; Stages 4–5 *see* post-construction phase
online survey tools 250–252
open procedure 159
operation management 239–243
operational policies document 223–224
operational reviews 253–254
organisational development 40
organisational skills 25, *26*
Out of the Crisis 6
outline planning permission 108
outsourcing 224

package deal and turnkey 148
parcels of work 67–69
partial completion 221, 222
participative leadership 27
partnering 153–154, 274g–275g
partnerships, innovative 162
party wall issues 111–112
PAS (Publically Available Specification) 1192-2 242, 243, 275g
PAS (Publically Available Specification) 1192-3 243
Pascale, Richard 38
patent defects 221–222
'pay when paid' clauses 194–196, 204–205
payee-led payment process 195–196
payment notices 195
payments 67, 194, 195–198, 203–205, 214
performance bonds 214
performance measurement 243–247
performance testing 236
performance-in-use assessments 254–256
PESTLE or PEST analysis (political, economic, social, technological, legal and environmental) 47–48, *48*
Peters, Tom 38
physiological needs **31**
plan–do–check–act (PDCA) cycle 187, *187*
planned preventative costs 126
planning 49–51, 109–110
planning permission: overview **86–87**, 107–109; Building Regulations 110–111; disability legislation 112–113; Enterprise Zones 109; party wall issues 111–112; rights of light 112; taking possession 225
PMI (Project Management Institute) 11
police 82
Pont, E.I. du 8
portfolio management 4
post-construction phase: commissioning 234–239; end of contract report 229–230; facilities / data migration 225–227; facilities management 230–233; handover and operation 227–228; operation and maintenance 239–243; performance measurement 243–247; post-project review

228–229; practical completion 218–223; quantity surveyor 79; taking possession 223–225
post-occupancy evaluation (POE) 233, 243–247
post-project review 228–229
practical completion 199, 218–223, 263–264, 275g
pre-construction phase: approaches to 70–72; clients' role **75**; cost advice / whole life costs 121–129; design 99–107; planning permission and legislation 107–121; preparation 82–87; preparation and briefing 87–96; procurement routes 145–170; procurement strategies 143–145; project team members 74–82; RIBA Plan of Work (2013) 72–73; risk 129–142; Task Bar 8 73; value management 96–99
pre-contract plans **50**
pre-handover 233
preparation phase, clients' role **75**
pre-qualification questionnaires 151–152
Prevention of Corruption Act 1906, 1916 16–17
pricing 83
PRIME 44
prime contracting 154–155
PRINCE2 (PRojects IN a Controlled Environment) 41–44, *42*, 275g
The Principles of Scientific Management 5
prior information notices (PIN) 163
private finance initiative (PFI) 167
pro-formas 263–266
proactive maintenance 127
processes: evaluation of 255; variable and fixed **68**, 69
ProCure22 (P22) 167–168
procurement: competition 145; contractor designed portion (CDP) 150–151; cost advice 170–178; cost reimbursement contracts 150; design and build (D&B) 146–148; design management and 99–101; drivers *143*; Environmental Impact Assessment (EIA) 165–166; EU public procurement 156–165; management procurement 148–150; negotiated contracts 153; negotiation 144–145; new models of 156; partnering and frameworks 153–156; pre-qualification questionnaires 151–152; public–private partnerships (PPPs) 166–168; risk and 136, *144*, 170; routes 143; single-stage selective tendering 145–146; strategies 143–145; sustainable 169; target costs 153; term contracts / schedule of rates 153; two-stage competitive tendering 146
professional bodies 10–12, 76
professional indemnity insurance 65, 84, 85, 213
Programme Evaluation Review Technique (PERT) 8–9, 50
programme management 4
project audits 248–253
Project Information Model (PIM) 243
project management: definitions 2–3; development of 5–7; governance and professional bodies 10–12; life cycle *36*; multiple projects 67–69; phases of 45–52; pitfalls 45–46; the role 21–24, *54*, 74–76; skills 1, 24–36; terms of engagement / appointment 83–84; timeline 8–10; tools and techniques 41–45; variable and fixed processes **68**, 69
Project Management Body of Knowledge (PMBOK) 9
Project Management Institute (PMI) 11

project managers 74–76, 85
project meetings 191
project monitor 3
project reviews 254
project sponsors 74, 275g
project team members: choice of 73; construction phase 179–182; roles of 74–82
projects 1–2, *25*
PRojects IN a Controlled Environment (PRINCE2) 41–44, *42*, 275g
prompt payment code 198
Property Advisors to the Civil Estate (PACE) 96
Public Bodies Corrupt Practices Act 1889 16
public–private partnerships (PPPs) 160, 166–168, *168*
public procurement 156–165
Public Procurement Directives 157, 158

qualitative risk analysis 133, 141
quality assurance (QA) 183
quality audits 183–184
quality constraints 23
quality control (QC) 183
quality management 182–184
quantitative risk assessment 133–134
quantity surveyors 78–79, 182
questionnaires: post-occupancy evaluation (POE) 249–252; pre-qualification for contractors 151–152; sample 269

reactive costs 127
reflection 16
reputation 19–20
reserved matters 108
restricted procedures 159–160
retention: bonds 214; funds 225; release of 196
RIBA (Royal Institute of British Architects) 76
RIBA Plan of Work (2013): overview **71**, 72–73; BIM and 102–104; classic project management stages and **47**; design responsibility **107**; order of cost estimating *174*; Soft Landings **86–87**, 230–231; Stages 0–4 82, 87; *see also* pre-construction phase; Stage 5 *see* construction phase; Stage 6 *see* post-construction phase; Stage 7 *see* occupancy phase
RICS (Royal Institution of Chartered Surveyors) 11–12
RICS Building Cost Information Service 123–124
RICS New Rules of Measurement NRM 1 89, 172–178
RICS New Rules of Measurement NRM 2 89, 125–129, 241–242
risk: accountability 130–131; allowance estimate 176–178; analysis 140–142; assessment 133–134; attitudes to 132; avoidance 131, 135; constraints 23; contingency 131; costs and 123–124; dealing with 131–132; definition 129; identification 132, 137–140; management 46, 132; mitigation 131–132; monitoring and control 134; procurement 136, *144*, 170; reduction 135; register/list 138, **139**, 176–178, 267–268; response 134; retention 135–136; take no action 132; transfer 132, 135–136
robots 66
Royal Incorporation of Architects in Scotland 76
Royal Institute of British Architects (RIBA) 76
Royal Institution of Chartered Surveyors (RICS) 11–12
running costs 124

safety needs **31**
SAVE (International Society of American Value Engineers) 96, *98*, *100*
schedules: for handover 227–228; of rates 153; of services 83; unrealistic 45
scope of projects 22–23, 45–46
Scottish Futures Trust (SFT) 167
sectional completion 221, 222
self-actualisation **31**
sensitivity analysis 142
sequestration 201–204
single-stage selective tendering 145–146, 170
site meetings 190–191
site works 189–193, 230
skills: BIM 91; leadership 26–29; project management 1, 24–26
snagging lists 222
social needs **31**
Soft Landings 84–85, **86–87**, 230–232, 275g
software packages 66, 89–92, 140–141
space planning 238
Space Shuttle Challenger project 9
sponsors 74, 275g
staff attraction / retention 19
stage payments 197, 204–205
stakeholders 257, 275g
Standard Form of Cost Analysis (SFCA) 173, 175
standards of service 16
statutory approvals **86–87**
Strategic Advisory Group on the Environment (SAGE) 188
strategic brief 87–88
Strategic Definition 78, 82–83, 102
strategic reviews 256
structural engineers 79, 181
subcontractors 81, 216–217
success criteria 249
suppliers 81
supply chain management 5–6, 205–207, *206*, 209–212
supply chains *180*, 195, 207–209, *208*, 275g–276g
suspension: of payments 203–204; of performance 196
sustainability 92–96, 169, 244
Sweett Group plc 18
SWOT analysis (strengths, weaknesses, opportunities, and threats) 46–47, *48*
systems migrations 226

taking possession 223–225
target costs 153
task bars 72–73
Taylor, Frederick 5–6
technical design 79, 103–104
technical skills 25, *26*
technical specifications 164–165
tender bonds 214–215
tendering 145–146, 158–160, 164–165
term contracts 153
terms of engagement / appointment 83–84
time constraints 23
Town and Country Planning Act 1990 107
TraderTransferTrust 67
transparency 16
trickle migrations 225–226
trust 16
two-stage competitive tendering 146

UK Government Information Exchange 73
unethical behaviours 18
Usable Buildings Trust (UBT) 231, 260–261
utility supplies 225

value analysis 97
value chains 208

value engineering / management 5–6, 96–99, *98*, *106*, 276g
visual surveys 259

warrant officers 82
warranties 65, 215–217
Waterman, Robert 38
whole life costs 121–129, *122*
withholding notices 195
workers' CDM responsibilities **119**
working hours 200

Xerox 245